DISRUPTING INJUSTICE

NAVIGATING CRITICAL MOMENTS IN THE CLASSROOM

Lateefah Id-Deen and **Esther Song**

Foreword by **José Vilson**

A Publication of

NCTM | NATIONAL COUNCIL OF TEACHERS OF MATHEMATICS

1906 ASSOCIATION DRIVE | RESTON, VA 20191-1502

Director of Publishing and Creative Services: Scott Rodgerson
Production Manager: Sandy Jones
Copy Editor: Elizabeth Adler
DEI Reviewer: Dr. Jennifer Howze-Owens
Publications Coordinator: Mary Donovan
Cover and Interior Design: Kirsten Ankers
Layout: KnowledgeWorks Global Ltd.

The National Council of Teachers of Mathematics advocates for high-quality mathematics teaching and learning for each and every student.

Recommend citation:
Id-Deen, L. & Song, E. (2023). *Disrupting justice: Navigating Critical Moments in the Classroom*. National Council of Teachers of Mathematics.

ISBN: 978-1-68054-021-5
ISBN (ebook): 978-1-68054-022-2
Library of Congress Control Number: 2023945412
Cataloging in Publication data is on file with the Library of Congress.

Printed in the United States of America

Table of Contents

Foreword

When we say "math in the real world," what conditions do we assume to be true about our classrooms today? Too much of what passes for "real world math" is just a reification of the world we live in and, in many ways, seek to deconstruct. By deconstruction, I don't mean to obliterate. I mean, while we're building something new, we should carefully consider every brick that built the old institutions and figure out how we got here in the first place. Once we recognize the patterns that built those structures, we can build something that eventually gets us to institutions where we don't have to ask ourselves why these structures exist as they do.

Unfortunately, asking us to reconsider these structures is a threat to the status quo and those who uphold it. As my friend Sabrina Joy Stevens has communicated time and again, the systems are only as dangerous as the people who allow them to continue in perpetuity. As such, teachers who seek to disrupt the status quo also have vast opportunities to change structures by organizing peers and schools with transformation and justice in mind. Educators—this includes any number of children-facing adults—seeking to build this path forward understand the task at hand, including the plethora of risks involved. These include, but are not limited to: threatening wayward observations, assigning them the "bad" kids or classes (read: our students most in need of great teachers/teaching), not providing professional development and advancement, and yes, harassment, ostracization, and termination.

Don't ask me how I know.

We have a plethora of structures worth deconstructing, too. Schools and those who have power over them pay more attention to standardized test scores and compliance with data binders rather than clean drinking water and good classroom resources. Schools and those who have power within them still overly suspend children who our society has deemed undesirable and use other racialized children as rationale for our deficit model thinking. We have a plethora of frameworks and rubrics determining our interactions and behaviors without deep appreciation for context, history, truth, reconciliation, and, ultimately, repair. Our sites of schooling want a no-excuses culture without responsibility for generations of harm done to our students and communities more generally.

What's more, we must deconstruct these elements in real time and in real life. While we reflect on these elements, we still need to put in the work. While hundreds of thousands of educators have left, millions more have stayed to pick up the fragile pieces left behind. Waves of a global pandemic, social and economic stratification, racial uprisings and the ensuing backlash including a governmental insurrection, and mass shootings have plagued a country founded on slavery, genocide, and colonization. History continues to live through what we call reality, one class period at a time. While the powerful push for normalcy, the few and dedicated—like those in this book—have decided to rethink normalcy and help reconstruct something new, hoping to take present and future generations along for the journey.

As a veteran educator, I spent my first year visiting my peers' classrooms, becoming a conglomerate of other educators' practices and perspectives. As my professional aspirations grew, I received opportunities to visit more classrooms. In each instance, I picked up practices that aligned with my dispositions on teaching and learning, including those that challenged pedagogies I came of age with. The more I grew into my own as a teacher, the more I saw how much of teaching is communal. No classroom—including mine—was perfect, but we math teachers understand the asymptomatic nature of this work. We keep approaching the axis of justice in this way because it guides our work, our hearts, and our minds.

To that end, students asked me if they would use a mathematical idea in real life only when they did not have access to the math and didn't see themselves in it. The context and rationale for *why* someone continually would ask for "real life math" are missing from the discourse. We continue to bend our math standards and practices to real life without specifically asking our students for input on our pedagogy and without reflecting on how our lessons can grant access to them, particularly those consistently pushed away from their own brilliance. Thus, our classes are *very* much "real life" because, upon completion of the schooling process, our students will encounter how often people in power determine the decisions our students get to make (or not make) with regularity and in perpetuity.

Educators who fully believe in the power of disrupting injustice are educators who help to create a reality that allows access for everyone to be their fullest, sustained, and most loving selves.

Sometimes, societal notions of disruption mean students, communities, and educators get treated as collateral for wayward notions of competition and excellence. Profiteers and politicians inject hateful narratives into the discourse in service of shaking the public trust in institutions that serve everyone. In doing so, they seek to privatize these institutions—and thus our interactions within them—for their personal, economic, and political gain. By contrast, this book seeks to elevate disruption as a means of noticing deleterious patterns of racism, sexism, homophobia, and other oppressions based on identity. It also asks how we can do better together. The authors and contributors dissuade readers from current notions of cancellations and wokeness and focus more on how we can pour nourishment into one another and ourselves in the service of a truly shared humanity. We can't do the latter without addressing the former. We get to live with ideas of abundance and greatness that seek the betterment of the collective and our individual role in it.

We're better for working alongside what the authors and contributors have compiled here. Join us in this real-world problem solving.

José Vilson
Educator, Executive Director of EduColor, Author of *This Is Not A Test: A New Narrative on Race, Class, and Education*

About the Authors

Dr. Lateefah Id-Deen is an associate professor of mathematics education in the Bagwell College of Education at Kennesaw State University. She earned her doctorate in curriculum, instruction, and teacher education from Michigan State University, with foci in mathematics education and urban education. She is a co-author of the book *Middle School Mathematics Lessons to Explore, Understand, and Respond to Social Injustice*. Additionally, she has written several articles and book chapters that examine students' perspectives on their experiences in mathematics classrooms and ways to support educators in hearing and developing practice in relation to students' expressed interests. She investigates social justice pedagogies and culturally responsive instructional practices that promote student-teacher relationships, affirm mathematics identities, and cultivate belongingness to support students' learning experiences in mathematics classrooms. She works alongside prospective and in-service teachers, curriculum developers, schools, and districts that want to incorporate culturally relevant and social justice mathematics instructional strategies in mathematics classrooms. Her work reflects her passion for creating equitable learning environments for students from underrepresented backgrounds in mathematics classrooms.

Esther Song is the director of strategic initiatives for professional development at Nth Education Partners, an educational consulting group focused on mathematics, identity, and social justice. With nine years of experience as a high school math teacher in the Chicago area, Esther has also served as the district mathematics manager for grades 6-12 in Chicago Public Schools. She holds a B.S. in mechanical engineering from the University of Illinois at Urbana-Champaign, where her passion for math education took root. Esther's experience includes serving as a director on the board of the Illinois Council of Teachers of Mathematics, a member of the Publishing Committee of NCTM, and on the editorial board of the APME 2018 Rehumanizing Mathematics for Black, Indigenous, and Latinx Students book. Her work revolves around rehumanizing math classrooms through student voice, identity, and teacher reflection, promoting inclusivity and equity.

Acknowledgments

We want to thank the contributors for their willingness to share their critical moments so teachers can learn about disrupting injustice by seeing examples, accessing resources, and reading personal insights from mathematics teachers. It is our hope that we shared your work thoughtfully and responsibly. Further, we thank the anonymous reviewers and Derrick Brooms, who provided thoughtful and thorough feedback on the chapters. Thank you to NCTM for publishing this important book, which adds to other publications that center equity in mathematics education.

Lateefah's Special Acknowledgements

I am thankful for my phenomenal family, mentors, colleagues, and close friends, whom I owe more than I can give. I will always remember your unwavering and consistent prayers and support. I want to thank mathematics teachers who work towards disrupting racial inequities that exist in mathematics classrooms. I am grateful to every mathematics teacher and administrator dedicated to improving the mathematics learning experiences that disrupt racial equities in the classroom and beyond. During early conversations about the book, Esther and I knew the importance of showcasing the ways research and practice inform each other, so I am grateful to the contributors who helped make our vision a reality. More importantly, to the students who are roses that continue to grow from concrete—I will not stop until you get the mathematics learning experience that you deserve!

Esther's Special Acknowledgments

I am especially grateful to the leadership team at Nepantla Teachers Community, Jerica, Chanel, and Jen, who have continued to push my thinking and provide space for conversations like these. Many thanks to Liz and Darlene, who've sharpened my mathematical thinking, and to Ray for his loving support. Lastly, I want to thank the person who started my own journey in this work and whose influence can likely be found throughout these chapters, Dr. Rochelle Gutiérrez.

It's hard to believe that this book exists, that our wonderful book contributors shared their stories, and that a few years ago it began with just an idea. When Lateefah and I started the writing process, we wanted to center math teachers' needs. In short, this book was made for teachers—for the teachers who read the books on social justice, rally with the causes, see injustices in the lives of their students, and must also prepare their lesson plan on triangle proofs the next day; for the teachers who believe in detracking and find secret relief in teaching their honors trig class; for the teachers who feel like they are making no dents in the waves and yet come to school every day with their best; and for the teachers who are brave, resilient, and tired. This book is made with love and respect for you.

The Contributors

The critical moments in this book describe how mathematics teachers disrupt injustice and display a wide range of experiences and perspectives. This section includes a short biography and a picture from each contributing author.

Tandrea Fulton is a doctoral candidate at the University of Central Florida. Her research interests include the mathematical identity of Black female prospective teachers.

Dr. Emily Bogusch teaches 12th-grade mathematics and college physics in Phoenix, Arizona. After teaching a newcomer math class for students with limited and/or interrupted formal schooling, Dr. Bogusch became intrigued by how sociocultural factors and past experiences influence learning. She subsequently earned a doctorate in educational psychology with a focus on learning throughout the lifespan. Dr. Bogusch has worked extensively to improve access to quality education for traditionally underserved students; this work led to her National Board Certification in English as a new language. Her current interests include integrating real-time data analysis and authentic scientific inquiry into her classes.

Evan M. Taylor is a middle school mathematics educator in Indianapolis Public Schools and a PhD student at IUPUI. He explores ways to embed ethnomathematics, social justice, and histories into K–8 mathematics learning spaces. His research interests are Black mathematics education experiences in the US, ethnomathematics, and culturally responsive mathematics teaching (CRMT).

Dr. Kendrick Savage is from Oxford, MS. His teaching career spans more than 16 years of teaching mathematics and mathematics education courses at the university, community college, and public high school levels. As an African American male, it is important to him to be an inspiration. He currently serves as an associate professor of mathematics at Georgia Gwinnett College. He is committed to increasing the mathematical motivation of every student he teaches by connecting the mathematical content to students' culture, experiences, and personal interest, and helping students see the power in mathematics as a mechanism to solve social injustices.

Dr. Tashana D. Howse is a professor of mathematics education at Georgia Gwinnett College (GGC). She has mathematical teaching experience ranging from classroom public education to developing the teachers of tomorrow through preservice teacher education courses. Howse teaches preservice teachers and supervises elementary and secondary interns. Her research interests include culturally responsive teaching, student engagement in mathematical practices, and the use of technology to enhance learning and teacher content knowledge. Dr. Howse received both a bachelor's and master's of science degree in mathematics education from Florida State University. She completed her doctor of philosophy degree in mathematics education from the University of Central Florida

Michelle Lo (she/her) is a high school mathematics teacher and continuous learner. As a recent graduate of the Stanford Teacher Education Program (STEP), she taught mathematics in San Jose for 2 years before moving to Los Angeles and is interested in leveraging education as a space for learning and liberation. In particular, Michelle is interested in teaching mathematics as a means of civic engagement and social justice while improving STEM equity for girls, students of color, and multilingual students. She hopes to continue learning from and with others in her community and growing her teaching practice.

Teresa Dunleavy I am a learner. I am an educator. I am a mathematician. I am a white, genderqueer woman. These identities influence how I have approached working with students, teachers, and teacher candidates in university and PK-12 classrooms for the last 20+ years. In that time, I have worked as a paraprofessional, high school mathematics teacher, coach, classroom substitute, classroom researcher, and university professor. Across this work, I interrogate systemic inequities and racism through practices that dismantle white supremacy and support the development of antiracist, justice-oriented, discourse-based teaching and learning communities that center students' humanity.

Introduction

Mathematics teachers can stand for hope and transformation, shaping the intellectual and emotional growth of their students. Within the intricate tapestry of mathematics education, there exist pivotal moments—some deliberate, others serendipitous—where individual teachers wield the power to create more just and equitable classrooms and schools. The critical moments described in this book, often hidden beneath the surface of daily pedagogy, have the potential to unravel long-standing inequities and ignite a flame of empowerment for all students.

This book embarks on a compelling journey, exploring the authentic experiences of mathematics teachers who have confronted injustices—moments that not only defined their professional journeys but also contributed to a more inclusive and equitable future. From diverse contexts and backgrounds, contributors invite us to witness the struggles, revelations, and breakthroughs that culminate in their quest for justice and equity in mathematics education. Accompanying these teachers' narratives, researchers contribute illuminating commentaries that situate each critical moment within research. Through their insightful perspectives, we come to understand how these individual instances connect to macrolevel inequities ingrained in our educational systems.

Chapter Structure

The subsequent chapters contain critical moments individual teachers have faced by describing instances when mathematics teachers intentionally create more just and equitable classrooms or schools. Each narrative, written from the perspective of a math teacher, is accompanied by a commentary or response from a researcher's perspective. This commentary helps readers place authors' narratives within the broader research body to identify the critical moments and their connections to macrolevel injustices. Further, the contributors provide resources, such as books, videos, and articles, which support classroom teachers and professional learning communities (PLCs) that face similar challenges, want to shift their practices, or learn more about the various ways to disrupt inequities in their respective contexts. Research and practice should have a relationship that informs each other. This book provides an authentic connection between research and practice that will help instruction and interactions in mathematics classrooms.

Synthesis of Contributor Chapters

Navigating Critical Moments: A Guide for Mathematics Educators to Disrupt Injustice includes six chapters in which contributors identify and expand on critical moments when they make difficult but brave decisions to disrupt unjust structures or attitudes in the classroom or school that affect students learning mathematics. In Chapters 1–2, we present a framework that helps the reader identify injustices, generally and in mathematics education at various levels and consider ways they can disrupt injustices in their context.

In Chapter 3, Tandrea Fulton shares a time when she provided a young Black male an opportunity to enroll in an advanced math class due to his strong mathematics competency. She describes how she developed his confidence in mathematics while realizing he was trying to persevere through a challenging home life. Her story reminds us there are many ways teachers can support Black male students in mathematics classrooms.

In Chapter 4, Esther Song and Lateefah Id-Deen illuminate how a teacher made a choice to discuss a student's negative comment about his mathematics ability due to his race. Esther used that moment to have a classroom discussion that disrupted negative stereotypes and affirmed Black students' ability to do mathematics.

In Chapter 5, Emily Bogush highlights an unexpected conversation during an equity-focused professional development workshop with a colleague who shared negative generalizations about his students. She concludes that, even if it is one interaction, there is always an opportunity to speak with colleagues, challenge harmful assumptions, and gain a deeper understanding of their beliefs and practices.

In Chapter 6, Evan Taylor critically reflects on his intersectional identity as a Black cisgender man who continually navigates patriarchy as he supports young Black girls in mathematics classrooms. He describes the importance of reflection, which can shape and reshape instructional practices, specifically his interactions with Black girls.

In Chapter 7, Kendrick Savage and Tashana Howse examine how an African American male's statement in his mathematics class not only revealed his invisibleness, but also revealed his white teacher's inability to address his racialized comment. They explain how an African American male teacher grappled with responding to this moment while realizing his own challenging experiences mirrored this student.

In Chapter 8, Michelle Lo and Teresa Dunleavy describe how a first-year teacher debated how to respond to a political statement made by a student during math class after another student contacted her to share the harm it caused. Students' voices and agency prompted the creation of the restorative circle regarding the harm inflicted on queer students and students of color in the class. The situation challenges the notion that mathematics and teaching are apolitical spaces and demonstrates the need to make room for courageous conversations.

Taken together, these six chapters are examples of critical moments that explain how teachers disrupt injustices. We conclude the book with templates that will help mathematics teachers craft their narratives, individually or collectively, and identify pivotal moments that hold the potential to disrupt injustices. We hope this book offers unique insights to amplify the importance of authentic connections between research and practice to describe and name injustices.

Connecting to Catalyzing Change

National Council of Teachers of Mathematics (NCTM) published the Catalyzing Change series across the K–12 grade bands. The purpose of these books is to ensure that stakeholders continually engage in critical conversations around key issues to make certain that our students have high-quality mathematics learning experiences.

One challenge that the Catalyzing Change series highlighted, which directly connects to the aims and scope of this book, is dismantling structural obstacles that stand as barriers to positive mathematics experiences for students (NCTM, 2018). The chapters in the Catalyzing Change series identify challenges that continue to perpetuate unjust structures for students from underrepresented backgrounds in their classrooms or school. The books named some of those structural obstacles, which include tracking students into qualitatively different or dead-end mathematics course pathways, with some students being denied access to engaging and rigorous mathematics, and teacher tracking that leads to increased isolation and burnout for early career teachers.

We are aware that dismantling structures to create more diverse, equitable, and inclusive mathematics learning environments is not an easy task. So, we focus on educators' journeys as they navigate the resistance that comes from moving toward systemic change. Ultimately, it is the responsibility of institutions and organizations to address inequalities that affect our students. However, we want educators to hear how they can claim their power and disrupt injustices.

Explanation of Key Words

We list additional helpful definitions and commentary about deliberate word choices below. We will refer to them throughout the book.

Table 1. *Key Words for Reference*

KEY WORD	DEFINITION
BIPOC	BIPOC refers to the acronym for Black, Indigenous, and people of color.
POC	POC refers to the acronym people of color.
White	Since 2020, media publications have addressed the title case in which the term is used. Some continue to use lowercase to acknowledge past anti-Blackness. Others have decided to use upper case. We have decided to use lowercase in this text (Daniszewski, 2020).
Racism	The act of directly or indirectly devaluing one or more ethnic or racial group while uplifting and privileging another (Battey & Levya, 2016).
Sexism	Ideas of females not being suited to do math dates back to 1970s research, although there are notable women as far back as the early 1700s. These ideas and stereotypes persist today (Brown & Stone, 2016).
Bias	Perceptions that are either in favor of or against a person's or group's ability and performance. Both teachers and students can have unfair perceptions (Copur-Genetrurk et al, 2019).
Privilege	Benefits or advantages afforded to a particular group based on societal dominance. These benefits may be subtle such as receiving a less severe consequence for the same behavior (Battey & Levya, 2016). For example, a Black student is given detention while a white student gets a verbal warning from the teacher.

Ultimately, the concepts listed above are all connected. Biases start as indetectable and automatic associations such as "Asians are good at math" and "girls are not logical thinkers" follow. When those biases remain unexamined, privileges are afforded for groups perceived as superior in the form of recommendations for acceleration, fewer consequences for negative behavior, feedback that shows a teacher believes in them, and countless others. In this book, we begin to uncover how racism, sexism, and other social constructs influence math classrooms and ourselves as educators. With this understanding, we invite mathematics educators to do the deep and meaningful work of disrupting injustice.

PART I

CHAPTER 1

Unmasking Inequities in Mathematics Education

At the core of this book is the embedded belief that all people are equally worthy of dignity and seen as human. When we refer to disrupting injustices, we mean actions that interrupt beliefs, policies, or structures that demonstrate that one group (e.g., race, gender, class) is inherently inferior or less deserving of human dignity or conversely that a group should be considered superior (e.g., white supremacy, patriarchy). Unfortunately, white supremacy and patriarchal constructs show up in mathematics education through cultural or media-driven narratives as well as the microlevel person-to-person or classroom interactions. There is a plethora of societal messages that elevate the groups perceived as superior as more worthy and intelligent. They are so frequent and normalized that they become invisible, akin to air pollution. For our purposes, we define these terms in Table 2 with an example within mathematics education on how they might appear.

Table 2. *Social Injustices in Mathematics Education*

TERM/DEFINITION	EXAMPLE IN MATH EDUCATION
White supremacy refers to a system that maintains white economic, legal, political, educational, and social privilege (Kendi, 2019).	Textbooks (or an internet search) do not recognize many mathematicians of color despite their crucial contributions to mathematics (Sleeter & Grant, 2017).
Patriarchy is a social and political system that gives men privileges, which results from a socialization process that teaches people heterosexual men are superior to women (hooks, 2004).	To prove they are capable and exceptions to the rule; women and girls feel they have to work independently on math projects (Sassler et al., 2017).

During a math department meeting in my first year of teaching, I (Esther) along with the rest of the math teachers at a racially diverse school near Chicago listened as our department head/supervisor shared the latest achievement data by race. He pointed to the gap between Black, white, and Asian students. The trend lines showed decreasing gaps for Black students at our sister school in the district. And though our own school's data showed that while everyone's scores were increasing, gaps between races remained the same.

One veteran teacher in a department of mostly white veteran teachers raised her hand and asked curiously, "Why do you think these gaps exist? Do we have a sense of the underlying reasons? Why are our Black students consistently at the bottom?" Other teachers, none of whom were Black, seemed equally interested in the answer. Our department head thought for a moment and responded, "I've noticed that my Black students have a culture that isn't helpful for academics. They're always trying to get each other to goof off and do other things instead." I looked around at the rest of the math department, waiting for someone to say something. Mostly I saw a few nods and people seemed ready to move on to the next agenda item.

It may seem like a typical mundane meeting, but it's one of many examples of how racism is embodied in our education system. When people think about racism in education, scenes of an old white teacher throwing derogatory racial slurs at a Black child might come to mind. It could also look like a Black third grader in handcuffs dragged out by police officers while a white teacher looks on approvingly. These clear examples of racism are real and harrowing. It can be easy, then, for math teachers to look at their own context and believe racism does not affect their content or classroom given no such explicit evidence. But there are many more subtle ways that racism is weaved into the fabric of mathematics education.

In the example of the math department meeting, we could unpack several layers of how racism affected that moment in historical, structural, and interpersonal ways. I surface some questions regarding them:

1. What historical laws and social influences have made it more difficult for Black students to receive high quality education? How have these affected Black populations in generations past and today?
2. What gatekeepers exist in mathematics that prevent POC from becoming math teachers? In a district that serves a population with majority students of color, why are 90% of the math teachers white?
3. Do teachers treat Black students with the same hope and expectation as students of their own race?
4. Why was it acceptable that the department head could treat Black students as a monolith? Would he have been able to say that white students have a culture of academic excellence and push one another to try harder without another teacher saying that we can't categorize all white students that way?
5. What kind of pressure would a Black math teacher have felt in a conversation like this?

These questions are not meant to be accusatory, but to serve as a lens for seeing how racism, patriarchy, and cultural bias can affect mathematics classrooms and professional interactions.

In the United States, racism is embedded both in structures at the macrolevel (institutional) like Questions 1 and 2 above, and in moment-by-moment interactions, questions asked, or even facial expressions at the microlevel (interpersonal) like in Questions 3–5. While mathematics educators may not immediately feel that they have the power to respond to macrolevel injustices, we contend that there are many opportunities to disrupt at the microlevel. Teachers can address microaggressions in various ways in the classroom.

Sue et al. (2007) define microaggressions as "brief and commonplace daily verbal, behavioral, or environmental indignities, whether intentional or unintentional, that communicate hostile, derogatory, or negative racial slights and insults toward persons of color" (p. 271). Disrupting microaggressions is essential and has long-term effects on how people of color view themselves.

Pause and Reflect

See Table 3 for some examples of microlevel and macrolevel injustices and the unintended messages they can have on students. Several of these injustices are supported by research. Citations are listed in case you want to learn more about that injustice. The list of examples in the table is not exhaustive. As you read through the examples below, consider whether these or others occur in your context and determine whether there are opportunities for disruption.

While many can assume that the work of teachers occurs just in their classrooms, their work is not small. The microlevel interactions described in this book can greatly impact how teachers engage with students and impact macrolevel change, which is drastically needed in education. Table 3 contains examples of various microlevel and macrolevel injustices and how they can happen in the classroom, with research citations to learn more about the injustices. Read through the examples and reflect on ways injustice can occur at the micro and macro levels.

Table 3. *Injustices at Macro and Micro Levels*

LEVEL	DESCRIPTION OF INJUSTICE	UNINTENDED MESSAGE
Macro	Historically marginalized students are over referred to and overidentified in subjective special education mathematics classes (Ford & Russo, 2016).	Misdiagnosing students with stigmatizing labels can negatively affect their self-esteem and achievement in mathematics.
Macro	Underrepresented historically marginalized students are not placed in gifted mathematics classes (Whiting & Ford, 2009).	Because of continual underrepresentation, it is believed that historically marginalized students are incapable of performing at a high level in mathematics.
Micro	Black students, especially Black boys, are more severely and frequently punished in the classroom than their peers (Woodward & Howard, 2015).	Black boys receiving more frequent and harsh punishment can damage students' perceptions of their school and negatively impact their academic success.
Micro	Teachers shorten a student's name that is perceived as too difficult to pronounce.	Names can have powerful meanings. This tells students they are not welcome unless they assimilate with the teacher's cultural and linguistic expectations.
Micro	Ask a student of color to represent their entire racial group when discussing sensitive topics during mathematics.	This assumes that people of color are a monolithic group and every person of color has the same lived experiences.
Micro	Give students of color backhanded compliments, such as "You are so articulate," or overly compliment them for comparatively simple computations.	These types of compliments send a signal that the students did not seem smart to begin with.

The narratives in this book empower and encourage mathematics teachers to disrupt injustice by first providing examples that include behind-the-scenes processing of mathematics teachers who have responded to these injustices in their mathematics classrooms. A commentary follows each narrative to offer the research perspective that situates significant themes within literature. Before we can move towards solutions, it is beneficial to name the injustice and get a better understanding of how it connects to larger systemic issues. Next, we include a list of resources to support classroom teachers who face similar challenges. Finally, we include guided reflection questions that can be answered individually or during a professional learning and development session to process how disrupting injustice looks in the reader's context.

The idea that mathematics has anything to do with race and racism is particularly difficult to believe. The myth is that math is the most neutral or unbiased subject. We offer further examples to give insight on how racism plays a role in mathematics education.

- Nationally and historically, education leaders have focused on achievement gaps in mathematics between Black and white students, often using white student achievement data (e.g., standardized testing, honors or Advanced Placement enrollment, etc.) as the benchmark for other racial groups. The underlying message is that the real learning standard is whiteness.
- Media images of people who are good at math are often Asian or white. And, of course, even with mathematics achievement of specific Asian American groups, they still are not the standard—thus, ignoring the higher performances of these students and firmly entrenching the privileges of whiteness.
- Mathematical prowess is used as a proxy for intelligence and access to coveted career tracks yet remains a gatekeeper for people of color in industries focused on data and computer science.

Even for teachers who are convinced that racism is a problem in mathematics education, they may believe that nothing can be done without changes from the top. In short, it's a job for policymakers, superintendents, or government officials who need to amend laws and dismantle oppressive structures. Teachers can feel helpless in the throes of political whiplash while waiting for change. After all, a single teacher's influence in national, state, or even their own school's policies can seem insignificant compared to what happens inside their classroom. Others may focus their attention to changing structures in their schools like creating revised recommendation processes for honors courses or detracking math classes altogether. These kinds of changes align more with addressing the macrolevel injustices described earlier.

However, the most difficult kind of change for teachers to make is in the frequent messages they send to students through almost imperceptible interactions (i.e., the microlevel interactions). For example, a Black student loudly exclaims, "Math is so boring!" or a Latinx student puts their head down to sleep through class. A simple move could be for the teacher to ask kindly, "How are you? Is everything okay?" It's easy to dismiss this response as obvious, but after the 6th student or the 3rd time, a response like this can be more difficult to sustain than any structural change. It challenges a deeply ingrained belief that these students are lazy and don't want to put in any effort. Truthfully, many teachers would reply to the question, "Do you think students are inherently lazy?" with an effortless handwave saying, "Of course not. I believe they need safe and creative spaces to grow." But when their student has nothing on their paper after 5 minutes and simply says, "I don't get it," that same teacher might begin with, "Did you even try?" rather than to smile knowingly and say, "What could help?" These moments, however, are the very building blocks of a safe classroom and warm community. They are the moments in action to see students as complex human beings and not caricatures of their race, gender, class, or home life. They are the moments that disrupt dehumanizing beliefs about who students are, what they are capable of, and their purpose in society.

Empowering Educators

Mathematics educators face the additional challenge of their content area being perceived as neutral or cultureless. Despite published works that point to the political nature of mathematics (e.g., Gutiérrez, 2013; Gutstein, 2006), topics like race, gender, oppression, and justice are seen as relevant only for social sciences or English class. A particularly motivated math teacher might insert a project analyzing social injustice through mathematics in between a packed curriculum filled with standards unrelated to the issues of today. But with pacing concerns and expectations for growth in standardized testing, math teachers may not feel like they have the time to prepare or facilitate these lessons.

It's not that there aren't teachers who have been successful in this work. Several examples of amazing programs, classrooms, and educators that truly build rehumanizing mathematical spaces but feel out of reach for the everyday educator already exist. Indeed, the stories in the book titled *Rehumanizing Mathematics for Black, Indigenous, and Latinx Students* (Goffney et al., 2018) are inspiring, filled with chapter after chapter of teachers who've partnered with their community for students to learn and love mathematics. Unsurprisingly, the educators in the text have worked hard to develop spaces, structures, and narratives that celebrate the strengths of students from underrepresented backgrounds in their schools or classrooms.

One of our intentions is to build the capacity for educators to take steps toward these kinds of efforts rather than feeling daunted by them. A mathematics educator who learns about social injustices and even develops a desire to act can still feel a sense of powerlessness enveloping them. An educator can attend social justice and mathematics education conferences, read books about racism in mathematics teaching and learning, and listen to inspiring speakers with calls to action. Still, the following Monday in front of students, it is difficult to know how to disrupt oppressive structures or transform the classroom. Although a multifaceted approach that includes structural and curricular reforms is important, racism and oppression are not limited to these structures. Math educators must realize that disrupting injustice is also necessary in daily and common social interactions.

We look to exhibit the internal processing that educators must grapple with instead of displaying the results of the implementation of a successful program. Traditional definitions of success hinge on if there is achievement growth on standardized tests and if there is an increase in the number of students who attend college, or some similar metrics. Here, we pose a different set of questions:

- How does an educator consider their own identity and how it affects their actions?
- What fears or risks does an educator assess when considering disrupting injustice?
- How does an educator respond to injustice given their context and social, political, or cultural influences?

It may seem counterintuitive, but we want to highlight the mathematics teachers' stories that may seem too insignificant to be noticed and spotlight the internal tensions they are experiencing.

These chapters provide real-life examples of mathematics teachers disrupting injustice. Moreover, we want to showcase how every mathematics educator can engage in action within their context. The purpose of this guide is to empower educators by sharing decisions and thought processes of educators that are rarely emphasized in professional development. These educators make a real, sometimes underappreciated, impact with their students or community. This book takes a deeper dive into the micro moments that reveal beliefs, biases, and our lived philosophy more honestly than any classroom structure, assessment, or curricular change can.

We offer no simple 5-step action plan to being a social justice math hero. Even if such a plan existed, each step would feel unachievable. We ask each reader to take a philosophical shift with us. Building on the work of Aguirre, et al. (2013) on identity, instead of simply focusing on the *what* (the lesson plans, the procedures, structures, or even the pedagogy), we also shift our attention to the *who* (the educator, their beliefs, and invisible influences). We assert that our identity is both in constant evolution and deeply rooted. We begin to truly confront ourselves in small, consistent, and difficult decisions that ultimately pave the path to the educator we hope to be.

Defining Critical Moments

Our deeper purpose is to encourage and sustain critical hope for change for mathematics educators of all kinds. We live in a society with injustices that have a domino effect over time and even across generations and can lead to unintended consequences. For example, a domino effect with tracking refers to the phenomenon where a students' initial placement in a specific mathematics track can significantly impact their mathematical trajectory, opportunities in the future, and self-perceptions. Therefore, it is essential for teachers and other stakeholders to be aware of the potential negative impacts tracking and other injustices can create for students.

All too often, passionate teachers ready to tackle the injustices of society burn out after a few years, feeling more hopeless than before. On the flip side, overly enthusiastic teachers can respond with cheery confidence that "everyone's a math person" to students who feel out of touch with the discipline. We need more than feelings of optimism with no action, which Duncan-Andrade (2009) describes as false hope. At the same time, we must be able to sustain for the depth of the challenge ahead.

Duncan-Andrade (2009) defines critical hope as teaching in ways that connect the outrage of young people to actions that relieve the suffering they experience in their schools and communities. For our context, critical hope lives in truly addressing unjust systems through meaningful reflection and dialogue with students and colleagues. The role of critical hope can play a central function in education because it acknowledges that there are systems that favor a dominant society at the expense of others. When we work towards dismantling unjust systems through meaningful and reflective dialogue, oftentimes disruption happens, and this is where critical hope can manifest and ultimately foster change and transformation in our mathematics instruction as well as with our students and colleagues.

We believe these narratives will give mathematics educators critical hope through examples of how they can begin to make moves toward justice today. Even more so, we want to encourage the educators that have been working moment by moment toward social justice but feel like they are making no progress. These stories are meant to be like a balm for the tired educator's soul. Lastly, we bolster the legitimacy of the importance of these narratives in disrupting injustice through a researcher's lens on how they relate to larger, systematic challenges. We remind educators that the path toward rehumanizing mathematics education happens one moment at a time, and they have the power to begin.

Critical hope described above sparks courage in critical moments. We define critical moments as moments of opportunities for impactful change. In other words, a time when an educator decides to make potentially difficult but brave decisions to disrupt unjust structures or attitudes in the classroom or school that affect students learning mathematics. A critical moment can be an event that explores the possibility of transformation or affirms the dominant/oppressive narrative. Transformation here refers to how the status quo is interrupted (individually or collectively). Interrupting the status quo provides an opportunity for change and transformation that can be extended beyond the moment it occurs for a greater impact.

Educators disrupting injustice must do more than recognize critical moments; they also need to move these critical moments into deeper levels of transformation. The critical moments described here showcase moments of courage and resistance that demonstrate an understanding of social justice. Furthermore, this book connects to pedagogical questions or moments and broader challenges associated with educators' identities and institutional barriers. The goal is to weave educators' perspectives and experiences and surface how disrupting injustices at the microlevel connect to macrolevel injustices.

CHAPTER 2

Disrupting Injustices: Conversations to Change Our Practice

Many people from social justice perspectives have used the phrase of doing the inner work, and sometimes that work can be difficult to navigate, particularly without guidance. But this inner work is a key part of resisting oppressive, dominant norms in our culture. It is, after all, impossible to disrupt injustice if we are unaware of it.

I (Esther) was in a grade-level meeting where we discussed what supports were needed for sophomores with failing grades. We split up into teams of four, and I was in a group with three white teachers. We talked about Malachi first, a Black male student who had stopped doing any work and only occasionally seemed to show up for class. The conversation felt somewhat hopeless. We couldn't do much for a student we didn't see often, and he didn't seem to care about his classes. We agreed that perhaps this was a situation for the counselor or dean to intervene. The next student we discussed was Ryan, one of the few white students in our school, who seemed more interested in sports and shoes than any academics. The white male teacher spoke up sympathetically, saying "I see myself in Ryan. I was a kid who got caught up in the wrong things and I just needed guidance to help me think through things. He's a good kid and maybe we need to approach him with a chat and some options for how he can make up missing assignments." The other teachers nodded in agreement.

I couldn't help but think that race had a part to play in our different reactions to the two boys. Sure, the students' situations and personalities were different, but the response to these boys seemed to be opposite when it came to the hopefulness of their future. I chose not to say something in response at that time because I didn't think there would be a productive outcome and I had a feeling I'd receive a negative response from the teacher. Nobody wants to be called out on being racist, and I could already hear the excuses forming. Inner work is difficult because we are being asked to identify injustice in society as well as the injustices we perpetuate ourselves. It's no easy feat to look in the mirror and see with sober judgment that we are part of the problem.

Responding to and disrupting injustice isn't always as simple as right or wrong, and one-stop-shop solutions are hardly adequate to combat hundreds of years of programmed beliefs. In my own city of Chicago, I watched a mayor proudly erect a Whole Foods, an expensive organic grocery store, in the middle of a low-income neighborhood. Seven years later, in 2022, it was announced that the location would close due to low sales. Is the solution to food deserts to erect more grocery stores in the area? Should the upper middle class stop moving into gentrifying neighborhoods? Is detracking the solution to inequitable access to content? Without a more contextualized, granular approach, it's unlikely that such solutions will produce results that honor marginalized populations with dignity and respect.

An 8th-grade algebra teacher friend shared her experience of how her district unilaterally decided to detrack classes in the name of justice. A few months later, she told me how teachers, unsure of how to account for the large variance in prerequisite skills, began teaching different content to three separate groups at different levels in one class. The policy had changed, but the people in the system had not. I'm sure there are countless examples of well-intended school or district mandates to increase equity or advance social justice that fell short in practice. It's not enough to adhere to a new policy reform. Disrupting injustice also requires attention to detail to interpersonal interactions, the beliefs teachers hold, and the actions they ultimately take.

An Approach to Facing and Responding to Injustice

We offer a four-section guideline template to identify and respond to injustices both inside and outside of the classroom. This guide offers an invitation to practice the skills of identifying injustice, reflecting on the social, political, and cultural influences that have shaped our own perspective, and responding with empathy. Accompanying every contributor chapter, it is meant to provide structure for the kind of difficult inner work specific to each teacher's context. We outline the sections below.

Section 1: Analyze a Narrative

It can be difficult for teachers to think directly about how injustice affects their classroom and school. Each chapter introduces a real narrative to ground the reader with a specific, relatable situation and provide a starting point to consider their practices. Teachers can leverage these stories to begin engaging with challenging topics like racism, sexism, or other power dynamics that come up in mathematics education without putting their own experiences on the line. This happens from a safe distance of another person's perspective with the intent of asking questions about how the author chose to approach the situation. We also invite the reader to consider how they would think or feel if they were in the same position.

Section 1 Guiding Questions:

- Where did the author experience conflict, internal and/or external?
- What aspects of the narrative resonated with you?
- What fears did the author share when considering how they would respond to the injustice?

Section 2: Weigh the Balance

Not every injustice can be met with picketing or demanding change from a supervisor. Teachers must also think of their capacity and the consequences they may face. In Section 2, we ask our readers to pause and consider what's at stake. We don't need to make snap decisions based on tradition, fear, or oblivious idealism. We do, however, need to understand our underlying motivations and concerns. In the end, disrupting the status quo comes with risks. Teachers must carefully weigh possible costs while they think about how they can respond to injustice.

Section 2 Guiding Questions:

- What fears do you have when addressing a similar moment? What negative consequences could result?
- Read the research section of the chapter that connects to the larger issues in education. What is at stake in choosing not to engage or respond?

Section 3: Consider Your Context

Typically, teacher professional development programs claim that their methods, programs, or curricula are scalable across a multitude of contexts. A good idea here will be a good idea everywhere. Although this is sensible for factories or producing goods, disrupting injustice at the classroom level is incredibly context dependent. Teachers have suffered through all kinds of professional development presentations claiming to be a game changer only to leave feeling like none of it applied to them. In this section, we pause to note the subtle (or large) differences between the chapter author and the reader. Teachers must consider their specific student population, identity markers, and resources when they witness or are perpetrators of social injustice. These differences mark opportunities to take action in a unique way.

Section 3 Guiding Questions:

- Have you encountered similar situations in your own practice? How were they similar and how are they different?
- How might the differences between you and the author change how you respond?
- Consider your identity markers, history, and cultural background. How do they differ from the contributor's?
- What assets can you uniquely bring to the situation? (E.g., personality, knowledge of the community, strong relationship with administration, race, class, access to university partners, etc.)

Section 4: Authentically Disrupt Injustice

When considering all the sections together, we don't hand out pre-made action plans. Rather, we give teachers space to create one that is authentic to their situation. We ask teachers to think about how they can still move forward considering their time limits, supervisors, curricular pacing, and school structures. Additionally, each teacher can harness unique assets, whether it's an ability to organize their peers, having a strong relationship with their administration, or even being easily likable. These very human considerations are often overlooked because of the inability to scale such attributes at large but ignoring them can mislead teachers into believing they are powerless.

Section 4 Guiding Questions:

- Responding to injustice in mathematics education often requires teachers to be vulnerable and brave. What are some possible risks that could be taken?
- How can you respond to a similar situation that is authentic to your context and yourself?
- What resources provided in the chapter could help build new learning that will help you?
- What do the next steps look like?

Table 4 summarizes the four sections and guides readers to use these chapters to disrupt injustice in mathematics education, centering the teacher's perspective. Chapters 3–8 will each include a tailored version of this guide for deeper processing and developing action plans.

Table 4. *Framework for Disrupting Microlevel Injustice*

SECTION	SUMMARY
Section 1: Analyzing a Narrative	Consider another's example of encountering injustice to empathize with a similar one of our own in a safe setting free from judgment.
Section 2: Weigh the Balance	Understand the injustice(s) encountered and pause to note the possible consequences, fears, and hopes in responding.
Section 3: Consider Your Context	Recognize that our personal experiences and identity distinguish our response from another's and understand why it matters.
Section 4: Authentically Disrupt Injustice	Utilize our unique characteristics and assets to respond to injustice with ingenuity.

The aim here is to help math teachers process the micro interactions in their professional spheres so they can respond to the injustices in their schools or context. We encourage teachers to engage with this guide in Chapters 3–8.

Guidance for Professional Learning and Development

This section is designed to assist individuals who want to collectively discuss and reflect on the critical moments shared in this book. Without reflection, experience alone might cause teachers to "reinforce stereotypes…offer simplistic solutions to complex problems and generalize inaccurately" (Ash & Clayton, 2009, p. 26). Reflecting with others provides opportunities for individuals to learn from not only their beliefs and experiences, but others that contribute to a wide range of perspectives. These reflections should take place in brave and safe spaces. We anticipate that people will utilize this book in learning spaces such as undergraduate or graduate university classrooms (equity focused or teacher education courses), professional learning communities (PLCs), or book studies held inside or outside of schools. In short, individuals can facilitate conversations about disrupting inequities in mathematics classrooms across various learning settings.

Shared spaces provide opportunities for individuals to problem solve with other educators and to learn strategies for engaging in practices and interactions that work to disrupt inequities. After reflecting individually, having conversations in a collective space could create opportunities for alternative perspectives and interpretations of the narratives and offer robust and thorough conversations that consider the reader's context, positionality, and background. We encourage teachers to discuss the narratives with colleagues, utilize the resources, try the strategies in their own contexts, and reflect on their experiences. Educators must work alongside one another to dismantle systems of inequities in mathematics classrooms and schools. Here are a few components to consider when facilitating a space to collectively reflect on the narratives shared in this book:

- **Cocreate goals for meaningful engagement.** Think about the big picture goals of collectively discussing the book as well as providing opportunities for participants to individually reflect on what they hope to learn from engaging with this book. Develop clear expectations and establish what you want participants to walk away with after your session(s). Determine goals that provide an opportunity to clarify your ideas and focus on your efforts, collectively and individually.
- **Establish group norms to create a safe space.** Conversations around equity and social justice require vulnerability, especially when it comes to reflecting on your own practice and history. For such conversations to flourish, they require great care in ensuring that participants feel safe to share their stories with one another. There are several resources that can help a group establish norms as they engage in discussions around equity. Some commonly used norms come from Singleton and Hays (2008), which are "stay engaged, expect to experience discomfort, speak your truth, and expect and accept a lack of closure" (p. 22). Staying engaged means to be fully present, intellectually and emotionally, during conversations. Expecting to experience discomfort reminds you that discussing topics, like systemic racism, can cause discomfort. Just remember that learning can occur on the other side of an uncomfortable conversation. Speaking your truth encourages you to be open and honest about your thoughts and feelings when talking with others. Expecting and accepting non-closure asks you to "hang out in uncertainty" (Singleton & Hays, 2008, p. 23) and not to rush to quick solutions without considering the nuances that exist.
- **Reflect on your positionality.** A person's positionality describes how their racial, gender, class, or other self-identifications, experiences, and privileges influence their engagement with equity conversations. To engage in conversations about how we can disrupt injustice, we start with what has influenced our own beliefs and perspectives. Embracing and reflecting on positionality encourages you to be aware of biases and privileges. In safe spaces, your positionality can be shared with the group to not only make others aware, but to help others know more about your perspective.

- **Make a plan to read and discuss the narratives.** There is no specific way to discuss the chapters in the book. This book is not designed to be read in a certain order. Decide how the participants will divide or select the chapters to read. We want to give the readers autonomy to engage with the book in a way that aligns with their respective contexts.
- **Consider time constraints.** Due to varying contexts, some groups will have sessions that range from 45–90 minutes although others will be able to hold consecutive hour-long sessions over time. As you previously read, this is difficult work, and it may take a lot of time to amplify the voices and perspectives shared in the space. Therefore, the coordinator leading the group should plan sessions that fit within your context while honoring the complexity of engaging in this work.
- **Complete the reflection templates individually.** In each chapter, you will find templates individuals can use for reflection after each narrative presented in this book. As you respond to the prompts, be completely honest and unfiltered. This is the time to surface unpopular or controversial beliefs and dissect your own beliefs. Additionally, completing the template before a group discussion allows for more opportunities to hear a variety of perspectives and take the time to deeply reflect on the narratives. It is not necessary to share every aspect of your individual work with others or a PLC.
- **Discuss and implement an action plan.** We want this book to inspire and empower teachers to disrupt their inequities in their own context. As you read the chapters, there is a wide range of what an action plan could look like, but more importantly how the micro moments can impact macrolevel issues. We ask you to visit Chapter 9 to determine what those moments could look like as you identify and reflect on critical moments that you already or may encounter in the future. This is where you can take your ideas and reflections and write solution-oriented, action-based approaches.

Recommendations for Interacting With Each Chapter

Ask the guiding questions after each narrative in the context of a conversation with others. Teachers are used to working in silos, dealing with their own burdens, and thinking of solutions on their own. We believe the narratives in this book have the power to bring teachers together through empathizing with similar situations and responding to the same injustice differently given their context and identity. We expect conversations in communities of math educators to spark deeper understanding of the injustices they see in their day-to-day experiences and yield truly transformational solutions.

We share our recommendations of how to dive into this work below. We write these to support engaging in the text and not as restrictions. There are many ways to interact with these chapters, and we want professional learning communities, facilitators, and individual readers to feel they have flexibility in how to engage with the text. Feel free to adapt these recommendations to your needs.

- Read the teacher's perspective first. This allows all participants to be grounded in a real encounter that people will likely find compelling.
- Discuss with a partner or a group Section 1: Analyzing a Narrative. Section 1 is used as a nonthreatening entry point for thinking about the narrative.
- Read the researcher's perspective to provide further considerations to better understand the injustices at play. Then engage in Section 2: Weigh the Balance. Participants can begin to discuss how the narrative connects with larger themes and the specific injustices that may need to be addressed.
- Provide individual time for Section 3: Consider Your Context so that participants can notice some similarities or differences between the chapter author and their own lives. It's critical that individuals have this time to understand how societal and historical influences shape them uniquely. It is the building block for the final section.

Alternatively, professional learning time with other educators can be maximized through prework. Participants can read the full chapter and write their own responses to each of the four sections before engaging in conversations around the questions. We provide templates for each chapter to provide space for written reflection.

PART 2

A Gateway, Not a Gatekeeper

TANDREA FULTON

I was a middle school 7th-grade mathematics teacher at a school where 70% of students received free or reduced lunch. The middle school was in the southeastern region of the United States in an urban area. It was a magnet school for the arts so it attracted students from all parts of the city. The school demographics were roughly evenly distributed among Black, white, and Hispanic students.

Recognizing the Potential

As the tardy bell rang and the last couple of students walked into the class, I said "Good morning, Asia, good morning, Daniel, and good morning, Maria." As I brought my attention to the board and started to ask students to do their bell work while closing the door, Malik somehow ran into the small opening of the door, seemingly without touching it. I turned, smiled, and said, "Good morning, Malik." Out of breath, he mumbled, "Good morning." One of our agreed-upon class expectations is that students come to class on time. Students who arrive late are given a visual cue that they will be marked tardy by the door being closed. Malik was often late or absent; however, he was always out of breath if he entered the class tardy. I respected his efforts and took it as a sign that he was at least running to try to make it to class on time.

I liked to incorporate a variety of bell work in my class. Sometimes the bell work would be one or two review questions from the previous lesson or the prior year, a problem analyzing a student's error, or "Which One Does Not Belong." On this day, the bell work was "Which One Does Not Belong," in which students had to analyze four expressions and determine which expression did not belong with the rest and explain and justify their selections. As I monitored the class to ensure that everyone was completing their bell work, I noticed Malik was looking around and did not have his bell work sheet. Without saying anything, I walked over to hand him a sheet of paper and a pencil. After a couple of minutes of giving students time to work, I instructed students to discuss their answers with their group mates.

When we discussed the bell work as a whole class, students were allowed to openly justify their answers. As one student finished saying their answer, another student would say "I disagree because..." or "I agree because...." Malik began by saying, "You could say all of them belong for some reason such as y^3, $2y + y$, $y(2 + 1)$ and $y + y + y$ all have [the letter] y, but I selected answer A because y^3 is $y \cdot y \cdot y$ and the others are the same as $y + y + y$." The students looked at me, and I looked back at the students and said, "What do you all think? Ask Malik questions if you feel you need further clarification." As they asked Malik questions and he explained simplifying expressions, I looked down at his paper and noticed it was blank. This was not the only time Malik showed a deeper understanding of mathematical concepts and that he could connect what we were learning to prior grades and other standards within the grade all while having a blank paper.

I went on to teach the lesson for that day and dismissed the class. The students left and I sat at my desk thinking about Malik. I knew that his family had been facing financial hardships and at some point were living in their car. I also often saw Malik walking his younger siblings to school, which was in the opposite direction of our school, when I was on my way to work. In the past, I had given Malik a book bag full of materials and hygiene products, trips to the school food pantry, and even shoes and clothes that were too small for my husband.

Processing the Conflict

I recently earned a master's degree in K–8 mathematics and science education. In those courses, we discussed and read articles about the mathematics pipeline and how there is an opportunity gap for minoritized students. I thought about Malik and his mathematical ability, and I felt he could be switched to my 7th-grade advanced math class period. However, I faced a dilemma. The 7th-grade advanced mathematics course moved at a fast pace and required students to do a lot of work at home. Missing even one day of class could put students significantly behind. I logged into my computer, pulled up Malik's grades, and saw that he had a 56% and many asterisks indicating several missing assignments. He also had 6 missing days, and it was only 2 months into the school year. I also clicked on scores and saw that he had no standardized mathematics test scores last year. I was faced with many questions. Is he interested in being in an advanced course? Do I request to change Malik to advanced 7th-grade math and risk adding more stress? Is it morally correct to decide what he is and is not capable of handling? I also began to wonder about the number of teachers who assess what they believe their students can and cannot handle and use that as a reason to not give their students opportunities. I decided to start by having a discussion with Malik after class the next day.

An Offer to Accelerate

The next day I asked Malik if he was interested in being in advanced 7th-grade math. He said, "Ain't I already failing this class?" I nodded my head yes and said that I thought he was more than capable. Then, I went on to explain the format of the advanced class. Malik and I had a lengthy discussion about how he perceived himself as a mathematics learner. Malik knew that he was good at math, and he, too, felt he could understand the concepts in an advanced mathematics course. However, he did not like doing work and did not care for homework. I asked why he did not complete the homework, as I assigned very few problems and even allowed students to get started on them in class if time permitted. He admitted that sometimes his house is noisy, so he cannot concentrate, or he just does not feel like doing it. He also professed that he misses a lot of days of school to watch his siblings when they are sick or if it is raining and they cannot walk to school. Malik went on to say that he stays 2 miles from the school, so there is not a bus that he can ride, and with walking his siblings to their school, he has to walk about 3 miles. In our school system, bus transportation is not provided if students stay 2 miles or fewer from the school.

I told Malik to go home and think about if he wanted to be switched to an advanced math class while considering the expectations of the class. I also went home and thought deeply about the articles and books I read during my master's program. I specifically remembered reading an article by Flores (2007) in which he discusses in his final comments that those who are considered to be on the top of the hierarchy of competence are given more opportunities to learn higher levels of mathematics and therefore have higher achievement and are perceived as more capable or having more aptitude. In contrast, those students who are placed at the bottom of the hierarchy are given fewer opportunities and are perceived as less capable. This perceived hierarchy of competence places whites at the top and Blacks at the bottom (Battey & Leyva, 2016). Thus, this notion of an achievement gap is partly due to an opportunity gap. I also recalled reading about early access to the mathematics pipeline and the implications it could have on students' interest in mathematics and selecting majors in mathematics-related fields in college. As a Black female teacher, I felt obligated to ensure that each and every student had the chance to engage in deep mathematics. Malik was a Black student, and I felt it was my responsibility to give Malik a chance at the early entrance to algebra. I also recollected reading counter-narratives written by Leonard and Martin (2013) and Berry et al. (2011), who highlighted success stories of Black males and their success in mathematics despite their circumstances and in light of all the beautiful dimensions of their cultures. This further cemented my decision of recommending Malik to an advanced mathematics class.

Connecting My Childhood Identity to My Teacher Identity

I am a Black female and I was raised in a relatively strict two-parent household by two Black parents in the southeast region of the United States. I was the second oldest of four and expected to have excellent behavior and good grades. My dad was a hardworking man, a truck driver who spent days on the road and nights at home. My dad dropped out of middle school to help care for his family. He thought education was important and the key to not having to work as hard as he had. My mom was a soft-spoken and very patient homemaker who prepared all our meals and made sure we were well taken care of while my dad was at work. She finished high school. My parents taught me that certain behaviors made you bad while other behaviors made you good. I wanted my parents and teachers to be proud of me, so I did my best to exhibit good behavior.

Hard work and good behavior placed me in accelerated mathematics in elementary school, which resulted in my taking algebra in the 8th grade. During this time, students that were on the standard track took algebra in the 9th grade, while many others took algebra in the 10th grade. Taking algebra in the 8th grade put me 1–2 years ahead of other students. This early access to algebra within the gifted program provided opportunities that other students may not have had including field trips, mathematics competitions, small class sizes, and highly qualified teachers. My parents did not accept excuses and expected the best. Though they did not attend college, the expectation was that my siblings and I would go to college.

I was a first-generation college student. I carried this trait to college and realized that others had many different experiences. I understood the things that I hated as a teenager, such as always having a parent at home to help me with my homework and parents who monitored my behavior, were a privilege that many did not have. I also realized that behaviors did not necessarily cause people to be bad. Still, perhaps it was a response to an environment, a coping mechanism, or a decision made from necessity. As I entered the field of mathematics education, I continued to become more open-minded about what intelligence and learning looks like and the teaching practices that are needed to provide all students with a high-quality education.

From Being Up for the Challenge to Feeling Challenged

The next time I saw Malik, he was excited to tell me that he wanted to try to be in advanced 7th-grade mathematics and would be willing to work extra hard to make sure he did well. I told him that I would be there to support him. After talking to his mom and pleading my case to the school counselor and assistant principal, Malik was placed in my 6th- period advanced 7th-grade mathematics class. Malik started strong; he completed and turned in his homework on time, finished his classwork, and maintained high classroom participation. He had a B average in the class and seemed to exude confidence and pride. This made me feel confident I made the right decision. I constantly told Malik that I was proud of him and that I always knew he could do it, and he would just respond with a smile.

In the second month of Malik being in the advanced class, he missed a whole week of school. I called home several times during this period and received no answer. I went from just checking on Malik with the hopes of him coming to school the next day to being concerned and worried by the fifth day. When Malik returned to school, he was not the same. I asked if everything was OK and if he needed to talk or go to the counselor's office. He said no. He said that his baby sister got sick and could not go to daycare, and then his brother got sick, so he had to stay home to take care of them while his mom was working. I told him that I tried to call to check on him and he said their phone was not in service. I continued to motivate Malik and explained that with a little extra work, we could get him caught up. Malik seemed defeated. I asked him if he was sure he was ok, and he said, "I really did not want to miss this much school. I knew it was going to put me way behind." I reassured Malik that we could get caught up if he was willing to put in the work. He agreed.

Over the next couple of weeks, Malik seemed to be off track. He would come to class late and did not have his homework or the makeup work from the days he missed. His participation diminished and he sometimes slept in class. I recommended coming to my class with his lunch during lunchtime to use it as a time to get caught up. He agreed but rarely showed up. Malik went from a B in the class to a D, and he was having behavior issues in other classes that caused him to be put in in school suspension, which resulted in more missed class days. During this period, I tried reaching out to his mom but still received no answer. I also talked to Malik after class to help get him motivated to complete makeup assignments and turn in class work. It was not working.

Questioning My Decision

During the last couple of weeks of school, Malik asked me what he needed to do to get a C. I printed out a list of his missing assignments. He completed just enough to get a C in the class. On the very last day of school, Malik stopped by my classroom on the way out and said, "Bye Ms. Fulton. See you next year." I said, "See you next year, Malik. Have a great safe summer." I sat back down at my desk looking at a room with bare walls, boxed books, and stacked chairs. I again asked myself similar questions to the ones I had asked myself months ago. Did I make the right decision by changing Malik to advanced math? Did changing his class negatively impact his mathematical identity? Was the format of the advanced class truly supportive of all students? Malik did not return to Whitfield Middle School the next year.

RESEARCHER RESPONSE **TANDREA FULTON**

Philosophy for Teaching

I am a Black female who has taught mathematics for 9 years at a middle school in which 70% of students receive free or reduced lunch. As a student who attended a school with a comparable socioeconomic status, I had similar experiences to the students I taught. I used my experiences and education coupled with the experiences and stories of my students to create a classroom environment that welcomes students of diverse backgrounds including varying cultures, ethnicities, histories, orientations, abilities, and understandings of the world around them. I pride myself in acknowledging who students are as individuals and finding ways to teach them through their experiences, perspectives, and cultures.

My teaching philosophy lies on a foundation similar to what Aguirre et al. (2013) mention in *The Impact of Identity in K-8 Mathematics Learning and Teaching: Rethinking Equity-Based Practices*. I believe that all students, in light of their humanity, including their backgrounds, culture, personal experiences, history, languages, and mental, physical, and emotional well-being, deserve opportunities to have a high-quality education that motivates them to make sense of mathematics and empowers them to use mathematics to promote change in the world (Aguirre et al., 2013). Therefore, my research focuses on the mathematical identity of Black male and female students and how teachers can improve their teaching practices to promote a positive mathematical identity for Black students.

Literature Supporting My Decision

In reflecting on my experience with Malik, I realize the impact that education and personal experience had on my decision to change Malik to the advanced course. Accelerating a student should be based on more than standardized test scores or a grade in the class—students showing a deep and coherent understanding of their current grade-level content through other avenues of assessment, such as explanations and justifications, is also worth considering (NCTM, 2020). When students are accelerated in middle school, they have early access to Algebra, and if completed successfully with a solid conceptual understanding, they are better prepared to take advanced mathematics courses in high school and improve their success in college-level courses. However, Black male students are underrepresented in advanced mathematics programs and courses.

There have been many reports about Black male students' low performance on standardized tests in comparison to their white male counterparts (Davis et al., 2019). Martin (2009) argues this so-called "racial achievement gap" supports the notion that some groups such as whites and Asians naturally possess a higher mathematical ability and represent the standard of competency (Davis et al., 2019). In contrast, Black male students are perceived as lacking mathematical ability, competency, or interest to learn mathematics. Berry (2008) provides research on eight Black boys who have experienced success in mathematics and describes five themes that emerged from the data. The five themes include early educational experiences, recognition of abilities, positive mathematical and academic identity, support systems, and alternative identities (Berry, 2008). As Malik's teacher, I recognized Malik's mathematical abilities and noticed he seemed to have a strong mathematical identity. I felt it was my obligation to allow him to take an advanced-level mathematics course as I supported him through the process. Most policymakers, administrators, and teachers in advanced mathematics are white and can serve as gatekeepers in Advanced Placement and gifted education (Davis et al, 2019). I wanted to be a gateway to give students opportunities to take advanced mathematics courses.

Deficit thinking and stereotyping about Black male students' abilities result in policies that are discriminatory and lead to low teacher expectations and poor-quality instruction that interferes with students' deep conceptual understanding and procedural fluency in mathematics (NCTM, 2020). Teachers not referring Black male students to advanced level courses, low standardized test scores, and low course grades are three factors commonly reported as reasons why Black males are underrepresented in advanced level courses (Berry, 2008; Davis et al., 2019). School administrators, teachers, and families can have an integral role in supporting Black male students in advanced-level courses. Families have been advocates for their Black boys to have access to gifted and Advanced Placement courses. Once gaining access, Black male students in advanced-level courses are often supported by their families through teaching, tutoring, and facilitating opportunities for deeper mathematical understanding.

Other Critical Moments to Consider

Retrospectively, I can think of missed opportunities with Malik and ways that I could have given him more support. First, I could have structured the advanced course differently to better support the needs of all students. Incorporating social justice tasks or culturally relevant mathematics tasks, such as using the case of Malik walking his siblings to school, could have gotten Malik and others more interested in the content. Among the things that impact students' learning experiences are the types and quality of mathematics tasks in which students are engaged, the discourse that is encouraged by the task, and the interactions that are fostered through successful implementation (Matthews et al., 2022). Culturally relevant mathematics can empower students by helping them make connections with themselves, their communities, and the world around them (Matthews et al., 2022). Culturally relevant mathematics promotes agency, which Gutstein (2003), as well as Matthews et al., (2022), describe as the way teachers and students view mathematics as a means of pursuing justice and equality in their communities. Agency is instrumental to a positive mathematical identity. At the time, I worked hard to stay on pace and cover all the 7th-grade standards as well as the required 8th-grade standards to prepare students for algebra the next year. Now, I realize that teaching 80% of the standards well with deep understanding is more impactful than rushing to teach 100% of the standards ineffectively.

Second, at the moment when Malik came back to school and I sensed something was wrong, I should have asked the counselor to talk to him. My gut was telling me something was not right, and because he did not tell me willingly, I let it go. Perhaps talking to the school counselor would have been beneficial for Malik. In addition, there was a club on campus called the Black Men of Excellence; I could have recommended Malik be a part of the club to have access to positive Black male role models. Black men have a unique set of experiences and circumstances in our society, many of which stem directly from the failures of key institutions that shape their development and provide opportunities for a promising future. In schools, for example, Black male students are faced with a larger number of referrals for school disciplinary action and are overrepresented in special education (Sánchez, et al., 2018). Similarly, in the criminal justice system, Black children are more likely to be sentenced as adults than white children. Given Black boys' experience with racism in the U.S and how various institutions negatively treat them, it is important to provide mentoring opportunities that help Black boys process these unique race and gendered experiences as well as teach them skills to effectively cope with these stressors (Hall, 2015). Malik having a mentor at school could have been another support in place to help him be motivated to stay on track.

I recognized that Malik had strong mathematics competency and provided him with an opportunity to take an advanced-level course. I gave Malik positive reinforcement and often told him I was proud of his hard work. These things appeared to have a positive impact on Malik's mathematical identity. However, there were opportunities to provide him with more support to help him get the best experience possible in the advanced mathematics course.

Resources

Tandrea's story highlights the importance of engaging students in tasks that pique students' interests by making connections to students' lived experiences and learning about injustices through mathematics. Additionally, Malik needed someone to believe in him and his mathematics ability, which can be finding programs that mentor young Black men and reminding yourself of the impact a teacher can make on a child's life. Below are resources that can be used to better support Black students' experiences in mathematics courses.

Table 5. *Suggested Resources for Supporting Black Students*

RESOURCE TYPE	RESOURCE	SUMMARY	PURPOSE
Book	*Engaging in Culturally Relevant Math Tasks: Fostering Hope in the Secondary Classrooms, Grades 6–12* by Lou Edward Matthews, Shelly M. Jones, and Yolanda A. Parker	A book that includes interactive and applicable tasks that novice and veteran teachers can use to shape and create culturally relevant experiences for all students while building standards-based lessons.	To support teachers who want to engage and empower students through implementing tasks that connect to students' communities and lived experiences.
Book(s)	*Catalyzing Change in Early Childhood and Elementary Mathematics, Catalyzing Change in Middle School Mathematics, and Catalyzing Change in High School Mathematics* published by NCTM	A series of three books (elementary, middle, and high school) explains the need for critical conversations to center around topics such as dismantling structural obstacles that stand in the way of mathematics working for every student and implementing equitable instructional practices to cultivate students' positive mathematical identity and a strong sense of agency.	To demonstrate the need for change and initiate critical conversations on policies, practices, and issues impacting mathematics teaching and learning.

(*continued*)

Table 5. *Suggested Resources for Supporting Black Students (continued)*

RESOURCE TYPE	RESOURCE	SUMMARY	PURPOSE
Book	*Choosing to See* by Pamela Seda and Kyndall Brown	A book that helps teachers examine their practices and beliefs and engage in transformative conversations around equity in the mathematics classroom.	To teach the equity framework I See You Care (ICUCARE), which includes a set of principles to provide all students with instruction that disrupts negative stereotypes of Black and Brown children. This framework embraces the ideology that doers of mathematics come in various ethnicities, races, and genders.
Video	*Every kid needs a champion* by Rita Pierson *https://youtu.be/SFnMTHhKdkw*	A TED Talk that explains why educators need to believe in their students.	To serve as a reminder on what it can look like to never give up on a student and empower them to believe in themselves.
Website	National Mentoring Resource Center https://nationalmentoringresourcecenter.org/resources-for-mentoring-programs/	Resources for mentoring programs including handbooks, curricula, manuals, etc., which can be used to develop mentor programs and incorporate practices in schools and classrooms.	To support creating or existing mentoring programs which provide opportunities for students to learn from individuals who can support their growth. Some students, like Malik, may have needs that require teachers to depend on others who may know how to support them outside of school.

REFLECTING FORWARD

Applying the Microlevel Injustice Framework

Use these guiding questions for your independent thinking and group conversations to move your thinking and actions forward around disrupting injustice.

<table>
<tr><td>Section 1:
Analyzing a Narrative</td><td>1. Where did Tandrea experience internal conflict and external conflict?</td></tr>
<tr><td>Section 2:
Weigh the Balance</td><td>2. What are the pros and cons of accelerating a student facing Malik's circumstances?
3. Read the research section of the chapter that connects to the larger issues in education. What is at stake in choosing not to engage or respond?</td></tr>
<tr><td>Section 3:
Consider Your Context</td><td>4. Consider how your context may differ from Tandrea's.
<table>
<tr><th></th><th>Tandrea</th><th>You</th></tr>
<tr><td>Race</td><td>Black</td><td></td></tr>
<tr><td>Gender</td><td>Female</td><td></td></tr>
<tr><td>Education</td><td>Masters in K–8 math and science education, read articles about opportunity gap</td><td></td></tr>
<tr><td>Years of teaching</td><td>9</td><td></td></tr>
<tr><td>School demographics</td><td>Black, white, and Hispanic. 70% of students receive free and reduced lunch</td><td></td></tr>
<tr><td>Teaching situation</td><td>Taught both regular and accelerated levels</td><td></td></tr>
</table>
5. Have you encountered similar situations in your practice? How were they similar?
6. How might the differences between you and the author change how you respond?</td></tr>
<tr><td>Section 4:
Authentically Disrupt Injustice</td><td>7. Responding to injustice in mathematics education often requires teachers to be vulnerable and brave. What are some risks you might need to take to provide acceleration opportunities for students at the bottom of the perceived hierarchy of competence?
8. What resources provided in the chapter could build new learning that will help you?</td></tr>
</table>

Who's Good at Math?

ESTHER SONG AND DR. LATEEFAH ID-DEEN

*I started the first class of the second semester in geometry with a new activity. I had spent all weekend creating a group game where students would respond to prompts and take turns guessing who wrote each prompt. Prompts ranged from "I learn best when..." to "The best thing about math is...." Groups enjoyed writing responses and laughing as they tried to guess who wrote them. As the bell rang, I told my students to turn in all their little slips of paper. I collected all of them and was about to put them in the trash when I decided I would take a moment to read through their answers. Some of the responses were funny, some were disappointing, and some were sad. Many students didn't feel like math was relevant, and most found it as something they had to get through to achieve what they wanted. One response to the prompt "Math is..." was especially painful to read. On the slip of paper, someone wrote "Math is... not for ni**as."*

Considering My Context

I was teaching in a selective-enrollment public high school on the south side of Chicago. A selective enrollment school is a type of public school that has specific admission criteria and a competitive application process. These schools are designed to provide specialized and rigorous education to high-achieving students.

As one of the few selective enrollment schools in Chicago Public Schools, students needed to test into the school, which touted a 99% graduation rate. With approximately 1,200 students, 70% identified as Black, approximately 25% identified as Latinx, and 65% of all students were eligible for free and reduced lunch. It was a unique place to be a teacher or a student.

In my master's program at the University of Illinois, I had the privilege to learn from Dr. Rochelle Gutiérrez through the Noyce Program as a professor, mentor, and now friend. The National Science Foundation's (NSF) Noyce Program, also known as the Robert Noyce Teacher Scholarship Program, is an initiative designed to address the critical need for highly qualified science, technology, engineering, and mathematics (STEM) teachers in K–12 schools in the United States. The program aims to recruit, train, and retain talented STEM professionals and undergraduate STEM majors to become effective educators in high-need school districts.

During Noyce seminars, our small group of preservice teachers began to unpack racism and oppression in mathematics education. With high hopes and a desire to make positive changes, I entered the teaching workforce. While I made continual attempts to improve as a classroom teacher for the first 5 years, I rarely discussed race, class, context, history, or students' lived experiences in the classroom. It's not that I didn't want to or that there wasn't an opportunity, but I preferred the safety of the mathematical learning standards or ideas surrounding character building.

By the middle of my 6th year of teaching, I was used to addressing inequity by working hard to increase access for students of color, particularly Black youth. I wasn't accustomed to addressing mathematical identity and its intersection with race. As I reread the response "Math is not for ni**as," I knew I could throw the paper away. No one would know that I had read it. Although the paper didn't have a name on it, I was fairly sure I knew who had written it based on the group—Jordan, a Black adolescent male who was bright and hilarious. I knew he had likely written it as a joke, and the rest of his group laughed when they read it.

I went home and spent the rest of the evening thinking about it. Questions upon questions arose. Should I bring it up with the whole class? Should I talk to the group? The student? What would I say? What would they say? Was it worth bringing up? It was, after all, meant to be a joke. But I also knew that part of it was not a joke but rooted in a student's reality. The school I worked at was unique in its location, programs, administration, etc. One of the school's few Black teachers said to me, "This is one of the few places you can be Black and smart at the same time." But even here, cultural beliefs were ingrained about which academic spaces were for Black people and which were not. If I didn't address it, no one would notice. We would continue life as usual, swimming along with the beliefs that mathematical intelligence and Blackness were antithetical. I wondered to myself, "Is this what I went into teaching for?"

My undergrad was in mechanical engineering at the University of Illinois. My parents came to the U.S. as Korean immigrants, and we lived in some challenging conditions. We eventually settled in the small section of affordable housing in a very affluent suburb. I spent most of my adolescence indoctrinated to desire this version of success. In middle school one weekend, I biked to a white classmate's house to hang out with some friends, and we spent our time in the carpeted basement with a pool table and arcade game machine. I didn't know life could be so luxurious. I returned home wishing I had lighter hair, a pointier nose, and a finished basement. I didn't start thinking any differently until my junior year in college. I attended an Asian American Christian conference that opened my eyes to the deep pains and surprising joys of being Asian American. My own slow transformation, one that continues to be ongoing, helped me to desire the same for others: a movement away from internalized racism and self-hatred and toward an appreciation for our backgrounds and differences from the dominant culture.

When I went to grad school to obtain a teaching certification, I began to learn more deeply about social injustice and reexamine mathematics education through an antiracism lens. Even with over 2 years of education on social justice and mathematics education and countless professional development workshops, some of which I had even led, when it came to the actual moment to disrupt oppressive narratives, I was afraid. The truth is, I didn't have much experience leading students in conversations about race and math. I didn't know how my class would react. But I had an opportunity, not a guarantee but a possibility, to reorient students' unchallenged beliefs of race and mathematical ability.

As a student teacher, I remembered having students working in groups on an assignment. I was making my rounds to the different hands that shot up asking for clarification. I redirected a Black girl back to her work and she looked at me and said loudly, "I don't do math. I'm Black!" It was one of several incidents in my career as a teacher where Black students, tired of a discipline that rejected them, pushed back that their race and mathematics were incompatible.

Crafting a Response

At home, I finally weighed the future outcomes of addressing this written comment. I had 2 other geometry classes and I hated the idea that I would be behind a day in one of them, a minor yet real concern as a teacher. As I wrestled back and forth, I finally took a big mental step back. Would students remember if we covered transformations for 5 days instead of 4 days? It seemed unlikely they would remember any of the 5 days by the time they were adults. But would they remember a moment when they addressed race in math class one time? The chances seemed higher. I decided it was worth the risk.

The next day, I began class by showing some of the responses to the whole class from the "Math is..." prompt. One student wrote "boring," another wrote "useful for my life," and someone wrote "interesting." Finally, I shared the last one, "Math is... not for ni**as." I held my breath. One moment passed in silence. And then, the class erupted in laughter. Internally, I panicked. What should I do? What should I say? I wasn't very surprised by the reaction, but I didn't have a plan. Jordan started laughing, pleased his prompt received so much positive attention. He lightly shrugged his shoulders and said, "I mean... am I wrong though?" More laughter. If only because I had no other recourse, I just didn't say anything until the laughter subsided. Their heads turned toward me looking for the next cue.

Tentatively, I asked, "What are your thoughts on this response?" Their faces got serious. As Black youth in a selective enrollment school, they were masters of the "school game." By school game, I mean that they have developed a high level of proficiency in navigating and succeeding within the educational system. While Black students may be good at navigating systemic inequalities, it highlights the resilience and strength of Black students in overcoming barriers while also calling for broader systemic changes to ensure equitable access to education and support for all students.

James, a Black student who had transferred midyear, raised his hand and said with exaggerated formality, "This isn't true. We should believe in ourselves." Several dignified heads nodded before giggling quietly. They got ready to move on, believing they had given the teacher what she wanted to hear. But, in my heart, I knew we were not ready to move on. I stood in silence because I wasn't sure what to say next. Caela, a high achieving Black girl and treasurer of the student board, said earnestly, "Wait, I'm not okay with this. Are y'all okay with it?" She looked at the rest of the class. "I'm not okay that to be Black means we can't be good at math." I appreciated her willingness to open up the floor and that she gave me some time to think. Jordan quickly replied, "It's just a joke. It's not serious." Several chimed in agreement.

Challenging the Narrative

I looked up and asked, "Is this the first time you've heard this kind of comment? That Black people can't do math?" I paused. Some students looked confused. It seemed like I was breaking the overrun professional script between teacher and student. I continued, "I'm asking, because it's not the first time I've heard a comment like this as a teacher." Caela was the first to reply, "I think this is something that society often pushes, in the media and everywhere—that Black people aren't smart." I agreed, "Yeah, it does seem to be an underlying stereotype. I mean look at me—I'm an Asian American teaching math." Students immediately laughed and the room atmosphere changed. The moment I put my own identity on the line, we stopped playing the teacher-student charade. We were a group of humans in a classroom discussing race and math.

James perked up, with a new openness this time, and said, "Asian people do be better at math." Internally, I knew this was not the conclusion I wanted students to leave with, but I also knew we were finally sharing our honest beliefs about race and mathematics. I had seen too many well-intentioned math teachers respond at critical moments like these with cheery, catchall remarks like, "That's not true, everyone is capable at math!" But statements like these didn't acknowledge students' experiences, and at worst, made them feel like they couldn't trust their own opinions. Black and Asian students are highly aware of the testing data and the gaps by race, and of who is sitting in AP calculus courses and who is sitting in remedial ones; they know the societal tropes that conjure up who is and is not a mathematician.

I paused and replied, "Why do you think that is?" James' friend, Kendall, spoke up, "Well, Asian people probably have quiet neighborhoods so they can study and actually do their work." James quickly responded, "Yeah, it's always so loud around me. I can barely get anything done or focus when I'm at home." Several students nodded, agreeing this could be a possible reason for the disparity. In response, I started sharing a bit about Asian American immigrant history and the socioeconomic status of Asians who were allowed visas, but I could see I was losing them. Their eyes started to glaze over. I pivoted and simply asked, "What are your thoughts now?"

Caela responded, "I don't want to agree that being Black is being bad at math. There are plenty of places that already say that Black people are stupid and that's not true." There were several honest nods this time. I could see our class period was nearly over, so I summarized, "Caela I think that makes a lot of sense. Even as a teacher I see this message repeated and believed often. But those aren't the messages we have to agree with." The bell rang and I sat back at my desk in the quiet. I was proud of myself for walking into my fear and finding a new appreciation for my students.

Reflections

I called Rochelle afterward and shared about the experience. She encouraged me to come back to the conversation and delve into the discussion further. I agreed to the idea, but I also wasn't sure what more I could say without some stronger planning on my end. I was already a day behind the other geometry class and I didn't quite feel I had the capacity to revisit the discussion. Even so, the following year James requested I write a letter of recommendation for him. I wrote fondly of his contributions to that discussion and felt a bit of regret that I hadn't pursued more conversations or spaces like those for him to shine.

RESEARCHER RESPONSE **DR. LATEEFAH ID-DEEN**

I am a Black woman who taught and worked with middle and high school students in urban and suburban contexts. As a student and teacher, I experienced and witnessed the inequalities in the education system. I helped increase middle and high school students' critical awareness through culturally relevant pedagogies through my teaching. As an associate professor, I support prospective teachers in discovering how instruction matters not only for students' development of mathematical knowledge but also for their sense of self. Further, my teaching style aligns with what bell hooks (1994) described as engaged pedagogy, whereby one teaches "in a manner that respects and cares for the souls of our students (p. 34)." Relatedly, my research and teaching centers on the importance of capturing students' perspectives on their experiences, cultivating belongingness and student-teacher relationships, and amplifying the importance of sharpening and enhancing students' mathematics identities.

First, I want to commend Esther for reflecting and then facilitating a conversation with her students about the problematic statement on the anonymous sheet of paper. The discussion provided an opportunity to capture a student's perspective. Students' perspectives on their experiences provide an important starting point for supporting their learning, especially in mathematics. Racism operates by systematically marginalizing underrepresented groups through interactions and practices (e.g., stereotypes and microaggressions), which materializes in policies (Sue et al., 2007). Not addressing or ignoring statements like the one the student wrote on the paper could lead to stereotype threat, which is mainly associated with Black students (Steele & Aronson, 2005). Steele (1997) defines stereotype threat as a risk associated with students from underrepresented backgrounds confirming to negative stereotypes that are often situational or incidental (p. 614). The statement that Jordan made can threaten how other students see themselves, altering various identities, engagement, and participation in mathematics. If these negative stereotypical beliefs about mathematical ability are upheld and fostered, students, particularly Black students, may begin to internally and externally disidentify with the school.

Stereotype threat can be difficult to interrupt in a classroom because it often goes unrecognized by teachers; however, there were a few important pedagogical maneuvers that Esther engaged in that helped to disrupt this injustice. With all the stereotypical voices spoken about Black students, schools—especially mathematics classrooms—should be the places that provide a counternarrative that Black students are creators and doers of mathematics. First, Esther amplified the voice of Caela, a high-achieving Black girl and treasurer of the school board, when she first spoke, which was a step in the right direction. To continue to build upon that moment, it would have been helpful for Esther to continue in future classes to showcase Black individuals who can do mathematics. Even further, this showcasing could have been reinforced with conversations about what it means to do mathematics. Too often, students' conceptions of mathematics are limited to school mathematics, which means successfully passing algebra, geometry, etc. Students need to see more examples of ways mathematics connects to their lives.

Second, Esther facilitated an honest conversation with students, which she did in her classroom. She acknowledged the racial stereotype and even raised awareness of her identity as an Asian woman who teaches mathematics. At that moment, she created a safe space for a difficult conversation that encouraged students to address their preconceived notions about race and mathematics. Creating space for students to offer feedback or share their perspectives about the class can open the class for new learning and sharing. Teachers might find a range of utility in identifying opportunities to raise students' awareness about societal messages or even stereotypes during lessons, whether from resources or the students themselves.

Unfortunately, Esther did not immediately revisit the conversation due to the fear of not following the pacing guide, like most educators. Hopefully, through Esther's story, educators can see the benefits of taking class time to facilitate critical conversations that directly impact students' mathematics learning experiences. While it may feel like a lot of time was lost, we must concentrate on what can be gained from having honest discussions with our students. Alternatively, an educator could find ways to have these types of conversations in more informal learning spaces.

When we reflect on situations, we often find missed opportunities we could have explored retrospectively. Therefore, I discuss another critical moment and explore why it was significant for Esther and her students. Earlier in the story, Esther mentioned that students wrote several responses, which included "Many students did not feel math was relevant, and most found it as something they had to get through to achieve what they wanted." These statements directly connect to students' mathematical identities. "Mathematical identity has to do with how one sees themselves as doers of, or learners of mathematics, the knowledge, skills, habits, attitudes, beliefs, and relationships that students need to develop as successful mathematics learners" (Aguirre et al., 2013, p. 14). Students' mathematical identities drive how they engage with mathematics and recognize and interpret their mathematical experiences.

When Esther saw statements related to students' feelings about math, there was an opportunity to recognize the importance of affirming and building students' mathematical identities. She did, however, acknowledge that she was not accustomed to addressing mathematics identity and race in her classroom. According to Martin (2012), mathematics teachers need to consider the interaction of racial identity and mathematics identity in children's experiences of learning school mathematics. Developing students' mathematics identities should be embedded in the daily work of teachers. Further, teachers can cultivate mathematics identities by leveraging students' culture and lived experiences, which empowers students to see the relevance of mathematics in their lives.

Resources

Esther and Lateefah highlight the importance of taking the time to disrupt negative stereotypes about who can do mathematics and present one way of engaging in a challenging discussion with students. Table 6 provides resources on cultivating a positive mathematics identity, navigating difficult conversations, and disrupting stereotype threats.

Table 6. *Suggested Resources for Fostering an Inclusive and Empowering Environment*

RESOURCE TYPE	RESOURCE	SUMMARY	PURPOSE
Book	*The Impact of Identity in K-8 Mathematics: Rethinking Equity-Based Practices* by Julia Aguirre, Karen Mayfield-Ingram, and Danny Martin	A book that helps teachers make connections between mathematics and student identities by describing five equity-based instructional practices that strengthen mathematical ability and mathematics identity development. It also provides information for how families and communities can assist students with their identity development.	To help teachers reflect and consider what is necessary to cultivate a positive mathematics identity, which can help students grow mathematically.
Website	www.mathematicallygiftedandblack.com	A website that highlights the contributions and lives of current and past Black scholars in the mathematical sciences.	To give students an opportunity to see aspects of themselves in Black scholars.
Guide	*Let's Talk! Discussing Race, Racism and Other Difficult Topics with Students* from Teaching Tolerance, a program of the Southern Poverty Law Center's project *Learning for Justice - Let's Talk Race*	A guide that has a wealth of information and provides free resources for educators and other stakeholders to discuss race, racism, and other difficult topics with students.	To impart strategies for engaging in difficult conversations and contribute resources that help teachers learn more about various topics.
Video	*Stereotype Threat: How It Affects Us and What We Can Do About It* by Claude M. Steele	A talk by Dr. Claude M. Steele about how stereotype threat affects people and what they can do about it.	To provide insight on ways educators can diminish the adverse effects of stereotype threat.
Article	*Countering Stereotype Threat* by Becki Cohn-Vargas	An article that helps teachers confront negative stereotypes that may occur in the classroom.	To share ways teachers can address negative stereotypes they hear in their classroom or see in other spaces.

REFLECTING FORWARD

Applying the Microlevel Injustice Framework

Use these guiding questions for your independent thinking and group conversations to move your thinking and actions forward around disrupting injustice.

<table>
<tr><td>Section 1:
Analyzing a Narrative</td><td>1. What aspects of Esther's narrative resonated with you?
2. What fears did Esther share when considering if and how she should respond to what Jordan had written?</td></tr>
<tr><td>Section 2:
Weigh the Balance</td><td>3. Read Lateefah's research section of the chapter that connects to the larger issues in education. What is at stake if a teacher chooses to not engage or respond?
4. What fears or challenges do you foresee if you needed to respond to a similar situation? What work do you need to do to overcome those fears and challenges?</td></tr>
<tr><td>Section 3:
Consider Your Context</td><td>5. Consider how your context may differ from Esther's.
<table>
<tr><th></th><th>Esther</th><th>You</th></tr>
<tr><td>Race</td><td>Asian American</td><td></td></tr>
<tr><td>School</td><td>Majority people of color, selective enrollment</td><td></td></tr>
<tr><td>Education</td><td>Learned about social justice in math education from an expert in the field</td><td></td></tr>
<tr><td>Years of teaching</td><td>6</td><td></td></tr>
</table>
6. What assets can you uniquely bring to a similar situation? (E.g., personality, knowledge of the community, strong relationship with administration, race, class, access to university partners, etc.)</td></tr>
<tr><td>Section 4:
Authentically Disrupt Injustice</td><td>7. When reading this account, what (if anything) felt risky about Esther's response? Where can you find space to take a calculated risk when responding to a similar situation?
8. What resources provided in the chapter could build new learning that will help you?</td></tr>
</table>

CHAPTER 5

Face-to-Face Encounters: Reframing Deficit Thinking as an Opportunity for (Shared) Growth

DR. EMILY B. BOGUSCH

Off-site professional development opportunities offer teachers a break from the day-to-day activities of teaching. For educators actively working toward a more just and equitable educational system, these opportunities can reenergize and reinspire us when we need it the most. But what happens when that highly anticipated workshop becomes yet another place to combat injustices? The first half of this chapter describes my experience in a professional development workshop that did not provide me with the relaxing escape and rejuvenation I thought I needed. In the second half of the chapter, I situate the experience within a sociocultural, research-driven perspective about teaching and learning.

Encountering a Colleague

Several years ago I attended a professional development workshop for integrating writing into the mathematics classroom. My state's department of education offers workshops that bring educators from across the state. I had already been experimenting with writing in my mathematics class, and I was eager to meet other like-minded teachers. This workshop presented writing as a gateway to more student-driven, cognitively engaging mathematics classrooms:

> Writing can provide opportunities for students to construct their own knowledge of mathematics…writing has given [students] a chance to practice inferring, communicating, symbolizing, organizing, interpreting, linking, explaining, planning, reflecting, and acting. Writing helps students make sense of mathematics. Mathematics helps students make sense of the world. (Countryman, 1992, pp. vi–v)

Like many math professional development opportunities at that time, the central premise of the morning was not writing, per se, but a more general truth: Everyone can learn mathematics. It was after spring break but long before that point in the school year when summer vacation seemed inevitable. I sat at a table with other teachers, but I only remember one. He was a first-year math teacher in a small school in a sovereign tribal community. The school is run by the education division of the tribal government and serves the children within their community. (Author's note: Identifying terms [i.e. sovereign tribe, tribal government] are those used on the official website of this community.)

I was in my 12th year of teaching mathematics in Arizona. I grew up in the Midwest and spent my high school years feeling jealous of my multicultural friends. At the time, I felt like my whiteness was the absence of culture, something long diluted from the process of immigration and assimilation. It felt like my friends had something that my life lacked: special foods, celebrations, and a heightened sense of community and belongingness.

Meanwhile, I already knew that I wanted to work toward a more socially just world, one where people celebrated difference, and everyone had the same opportunities that I had. Looking back, though, I now realize "social justice" encompasses so much more: social capital, inherited privilege, the distribution of wealth, access to health care, protection of civil rights, and so on. Years of personal and professional experience teaching in urban school districts has helped me to understand the depth of the term "social justice" while simultaneously reinforcing my commitment to societal reform. I realize both the limits of my own perspective—a white woman from a middle class, educated family—and the unique opportunities this identity provides me as an agent of social change.

Sharing My Views

Interspersed during the writing workshop were think-pair-share opportunities. I was paired with the new math teacher. At first, I was eager to learn more about the school and the community where my table partner worked, as well as his personal experiences in the past year. It quickly became clear that he was an outsider to his students' culture, and he was frustrated with his students and their class performance. One of our first think-pair-shares involved the premise that everyone can learn mathematics. His first year of teaching mathematics taught him that no, not everyone could learn mathematics. I challenged him by sharing my views that everyone could learn math—they might learn it in different ways or at different rates. Still, everyone could think mathematically, given the opportunity. My reflection partner acquiesced, and I felt like I had made a small impact on the world.

But during the next think-pair-share, he argued that although all students might be able to learn mathematics, the children attending this tribal school didn't want to learn mathematics. Key phrases were sprinkled throughout his explanation, and he made broad, negative generalizations about all his students. They didn't value school. They didn't care. They would use everything from cultural celebrations to funerals as an excuse not to attend school. When they were in school, they actively tried to avoid learning, and failing that, they tried to unlearn new skills or understandings if they were academic. According to him, the students he taught were uninterested in life outside the reservation; they planned to live their lives on the reservation and never strive for anything more. He explained how his students did not share in his values saying, "Personal improvement, formal education, and in particular, math, were irrelevant to their lives now and [in] the foreseeable future."

I politely disagreed, offered a counterargument or two, and shared my thoughts to whatever think-pair-share prompt we were supposed to be discussing. I don't remember the prompt or what I said, but I remember what I did next. It was an instinctual response, and not one that I am proud of. In the face of this anti-student negativity from a beginning teacher—a teacher with, potentially, a whole career and lifetime of influencing students in front of him—I followed my first instinct: I moved to a different table. I signed up for the workshop because I needed a break from my classroom. I wanted to be a student for the day, not a teacher. I wanted to learn new strategies to integrate writing. I wanted to share fun, inconsequential stories from the classroom with other teachers. I did not want to convince, persuade, enlighten, or educate. I didn't want to engage. I didn't want to keep talking to someone who was so negative—I encountered enough negativity from other teachers at my own school.

Continuing the Conversation

Sometimes we stumble into critical moments, and sometimes they stumble into us. The teacher followed me. He moved to my new table after the break and sought me out again after lunch to continue our talk about teaching and learning mathematics. As he moved his things to my new table, I felt my heart drop just a little. I looked around the room and saw other teachers smiling and laughing. I wanted to smile and laugh, too! At this point, I saw two options: move again or talk to him.

I decided to stay. I'd like to say it was due to my commitment to social justice, quality education, and students. Honestly, though, I stayed at that table and talked to him because I didn't want to be rude. Our society often views politeness as a weakness—it's the inability to stand up for yourself. And at the time, I felt like my professional respite had been hijacked by this persistent teacher. But after unpacking this encounter years later, I realize that politeness isn't weakness at all. Politeness is rooted in the belief that everyone deserves to be acknowledged, treated with respect, and heard.

Changing My Own Perspective

When I feel stuck in uncomfortable situations, I try to determine what I have control over. Since I had dismissed the idea of being overtly rude, I needed to find another way to shift the dynamic between myself and the teacher; otherwise, I was going to have a long and frustrating day. Instead of ignoring the teacher's negativity, or arguing at every turn, I decided to embrace this as an opportunity to affect positive change. I thought strategically about how to counter his negative viewpoints. I couldn't ignore him or his beliefs (I had already tried), but I didn't want to attack him, either. How could I confront his negative ideas—clearly linked to the culture and race of his students—in a positive, polite, and respectful way?

I asked several questions about the school, its students, and the larger community. He described parents who were nice but didn't care about education. They routinely pulled their kids out of school for weeks at a time for tribal and family gatherings like funerals or holidays. From his perspective, it was any excuse they can think of. I asked for specific examples that made him think his students were anti-learning. He did describe students that sounded defeated, students who were probably frustrated and disengaged from this particular teacher and his classroom. Next, I tried to pose leading questions to help him uncover his own biases: "How do you know?" "Why do you think...?" "How would you feel in their place?" Finally, after I felt I had laid the groundwork, I gave his point of view a name: deficit thinking.

Introducing My Colleague to a New Perspective

He had not heard the term before, so I explained it. I took some of his classroom examples and flipped the script: Just because middle school students don't bring pencils and books to class doesn't mean they don't care. Everyone wants to learn—there may be things in life we are not interested in, but when we gain new knowledge or understanding, we don't try to unlearn it. One of his counterexamples was that his students, on their way out of class, took the new pencils he gave them, broke them in half, and threw them on the ground. I asked him if maybe the students sensed his negativity and were reacting to him personally. "Oh, no. The students don't know what I think; I'm careful not to show it." My response: "They know." Teachers are under a microscope in the classroom, and students listen to what we say, how we say it, who we say it to, and who we don't say it to. They are masters at detecting the slightest hint of hypocrisy and double standards. Indeed, their very survival depends on determining which adults they can trust, and which ones they cannot.

We continued our conversation off and on through the rest of the think-pair-share activities. It was a little disjointed, but it gave him time to think about what I said, and to try to give me other examples that proved his point. It also gave me time to think, reflect, and strategize. I adopted what I thought was a nonconfrontational approach, in which I tried to question his assumptions instead of attacking his position. What didn't occur to me at the time, but I realize now, is that my position as a white woman meant this teacher would be *more* receptive to my message than if I was an educator of color. I was not his school administrator and not someone who had been marginalized in the educational system or disadvantaged by structural inequality. I looked like him, and that gave me credibility.

A More Promising Future for Us Both

What happened next? The workshop came to an end, and I started packing up my things. I was exhausted; I wanted to get home. Then, unexpectedly, my table partner walked up to me and thanked me! I had given him a lot to think about, and he was beginning to see his students differently. Or maybe he was finally seeing them for the first time.

Did I go home feeling self-satisfied and content? Maybe a little. But I was also seeing things differently. Why was my instinct to walk away from this critical moment? When have I walked away from opportunities to disrupt injustice in the past? How could I recognize those opportunities in the future? How did having the words—deficit thinking—help me to confront the issue? When can I find moments to step back and strategize conversations with my own colleagues? Finally, instead of waiting passively for a critical moment to follow me across the room, how can I create these critical moments—moments when change feels as inevitable as the start of summer vacation?

Since then, I have recognized these deficit-thinking opportunities are commonplace in the day-to-day workings of schools. While we may not have the comforting anonymity of a one-time professional development workshop, we always have an opportunity to engage fellow educators in a conversation, challenge harmful assumptions, and learn more about our own beliefs and practices in the process.

RESEARCHER RESPONSE **DR. EMILY B. BOGUSCH**

I have a doctorate in educational psychology; my research examines the intersection between culture, identity, and education. Broadly, how do culture, race, gender, and lived experiences shape the way people learn, grow, and teach others? How do we build on the strengths and resources of our students to create the educational experience students deserve? How can we support students in a way that elevates rather than limits? These are the types of questions that guide my own practice as a math teacher in a large urban high school and as a physics instructor at a nearby community college. They are the same questions I was grappling with when, years ago, I attended a professional development workshop on the integration of writing and math, a workshop where I tried to avoid a teacher that clearly needed a new way of thinking about his students.

Students Who Can't Versus Students Who Won't

Many of my colleagues believe one of the main equity issues in math education is the teachers who argue that some students can't do it, meaning these students can't do math, or rather can't do math at the level the teacher deems appropriate. This belief is one of the main drivers behind mathematics tracking (e.g., a calculus-track versus a less rigorous math course sequence). In reality, student tracking into different mathematical course sequences in high school is often correlated with "a variety of nonacademic factors, such as perceived (but not potential) academic ability, race, socioeconomic status, gender, language, or other expectations ascribed to students by adults" (Graham et al., 2018; Stiff, 2011).

The low expectations of many mathematics teachers working with traditionally underserved students is well-documented, as is the detrimental effect it has on students (Ladson-Billings, 2007; Valencia & Black, 2002). The approach of NCTM's Catalyzing Change series addresses the structural issues of ability grouping and tracking, but not necessarily how to confront this view directly with individual teachers. Instead, high expectations of all students is a starting point of literature outlining mathematics reform: a basic premise, helping to position the reader of such literature as an educator already dedicated to equity and inclusion (Association of State Supervisors of Mathematics et al., 2021). This assumption—that reform-minded educators have high expectations of all students—may be unproductive in the long run. In my encounter with the struggling first-year teacher at the professional development session, he quickly conceded the idea that his students couldn't learn math was an unfair judgment. However, attributing the failure of his students to their innate ability was a self-defense mechanism; the problem was outside of his direct control. Instead of shifting the responsibility to himself, he looked for other student characteristics to blame, and concluded that those particular students just didn't care. This conclusion is born out of the experiential evidence (i.e., lived experience) of the teacher, and thus it is more difficult for an outsider to challenge. Research into attributional theory and perceived personal control suggests that there are patterns to how people explain failure (Perry, 2003; Reyna & Weiner, 2001). The perceived cause of the failure will impact the perceived solution and its ultimate efficacy. Consequently, how students explain their own failure—and whether they can control it—may affect their learning just as much as innate ability (Dweck, 2007).

These explanations for academic failure are equally important from a teacher perspective. If students are not learning in the classroom, teachers have limited options for placing blame; the two most immediate options are either the teacher himself or the students. A study involving over 200 elementary school teachers in Cyprus examined teachers' attributions for student failure impacts their subsequent behavior:

> A student who faces serious difficulties with his achievement is more likely to receive better and persevering treatment by a teacher who has higher sense of efficacy, is willing to accept part of the responsibility for the student's failures, and sees the failing student as a challenge rather than a problem in class. (Georgiou et al., 2002, p. 583)

Just as students tend to be more academically successful when they focus on their personal control in learning rather than external factors, teachers will be more successful with students when they focus on their own pedagogical practice and the choices they make in their own classroom. In this particular study, the Cypriot teachers and students were both fairly homogenous groups with regards to race and ethnicity. Teacher explanations of student failure tend to differ when there is a mismatch between the culture, race, and gender identity of the students and teachers. For instance, a study of over 200 mentor teachers from Ohio (79% identified as Caucasian, 14% Black, and 7% other) posed a simple prompt: "Briefly describe an urban classroom" (Thomas-Alexander & Harper, 2017). The responses themselves indicated many teachers equated urban schools with low socioeconomic levels and large percentages of minority students. The researchers identified three basic responses:

> The first would be a focus on the strengths and opportunities available to those who teach and attend urban schools. These we classified as positive. A respondent may have chosen to focus on the problems and deficiencies of the students, teachers, parents, community, and physical plant. These we classified as negative. Finally, some respondents chose to offer a brief description statement that was neutral. (Thomas-Alexander & Harper, 2017, p. 54)

The responses directly reflected the teachers' perceived control in urban classrooms; descriptions of failure and low academic achievement were described as fixed and outside of the teacher's control, whereas positive descriptions emphasized ways in which teachers could utilize diversity to enrich educational opportunities.

Student Opposition

The power that teachers have in their own classrooms does not create a classroom culture in and of itself. Instead, the situational factors at play in the school and community matter, as does the agency of the students themselves. Students always have some agency in classrooms because they can resist and defy teacher expectations (McFarland, 2001). My colleague at the professional development workshop assumed his students were refusing education, and that this refusal was a byproduct of their cultural identity. An alternate explanation, which I suggested to him, was that the students were reacting to his personal beliefs and feelings. Likely this was exacerbated by specific classroom practices and the unique nature of mathematics education.

Hand (2010) describes this phenomenon in a low-track 8th-grade urban mathematics classroom. In this classroom, the teacher was a proponent of mathematics reform curriculum, had over 20 years of teaching experience, and was white. The students were predominately of African American and Latino descent (88% of the class, compared with a school average of 61%). Early in the school year, the teacher had attempted to engage students in a class discussion about their mathematical reasoning. Some students participated, but others were "not necessarily prepared academically to engage in these opportunities" (p. 106). By the middle of the year, the teacher attempted a similar discussion but with fewer opportunities to express their mathematical ideas. At this point, off-task behavior in the classroom was increasing. A few months later, she stated that "the interactions between the teacher and students had significantly deteriorated. The teacher provided fewer opportunities for the students to 'do mathematics,' and the ones that students received they actively resisted" (Hand, 2010, pp. 106–107). Her analysis highlights the way in which low student expectations can play out specifically in mathematics classrooms, rooted in a legacy of inequity:

> What is compelling about the features identified in this student is that they are related to the system of competence in the classroom, or what counted as productive mathematical participation...Tracked into opportunities to learn mathematics that negated their intellectual capacity and inhibited college entrance. I view the students' organization of an alternative system of competence a logical (albeit very detrimental) move against a system that marginalized them and their chances for success. From the teacher's perspective, however, the students' decline into unproductive behaviors perpetuated their academic failure, thus reproducing social and racial inequities. (Hand, 2010, p. 126)

Teaching students from underrepresented backgrounds poses a great many challenges, particularly when there is a cultural mismatch between teacher and students. The context of mathematics classrooms can further exacerbate this already tenuous dynamic.

Final Thoughts: Learning From Critical Conversations

When we engage in these critical conversations so essential to true reform in math education, we need to give colleagues the same consideration we give students: Allow others to learn from their mistakes. An assumption I made in this conversation was that the classroom dynamic the first-year teacher described was largely a product of his actions and his beliefs. In doing so, I made the same fundamental attribution error that he made with his students. The premise of the fundamental attribution error, as Ross (1977) originally describes, is that people "Overestimate the importance of personal or dispositional factors relative to environmental influence...[The intuitive practitioner] too readily infers broad personal dispositions and expects consistency in behavior or outcomes across widely disparate situations and contexts (p. 184). Just as teachers tend to attribute the success or failure of students to specific student characteristics, teachers also tend to attribute the success or failure of other teachers to personal or professional shortcomings (Kennedy, 2011). Education is inherently complicated, involving teachers, students, schools, communities, and a long history of inequities and power imbalances. We are all learning.

Recommendations for Action

So, what is an equity-minded, forward-thinking educator to do when these critical moments arise with colleagues? I have some suggestions, based on over 20 years of teaching experience and extensive academic research. Each of these suggestions has a rationale and a purpose; consider each one carefully and take the time to reflect on your own history of critical moments. To engage in equity work in mathematics education, it is essential that we think deeply about who we are, where we have come from, and how we engage others.

1. Assume positive intentions. Understanding students' thinking and motivation is at the core of what educators do. In Gilbert and Malone's (1995) dissection of correspondence bias—the mistake that we can infer the "character, motive, belief, desire and intention" that drives an individual's behavior—they delineate the process that people become convinced that they know why another person acted in a certain way (p. 21). The danger in thinking we can peer into the minds of our students is that we are often wrong. The problem is that we act upon these assumptions. When teachers assume students don't care about learning or aren't putting any effort toward a class, they change the way they interact with the student (Perry, 2003). Studies find that, consistently, "Teachers tend to reject the child and behave with anger toward him when the child's own inadequate effort is cited as the major factor for failure" (Georgiou et al., 2002, p. 593). If the teacher gives up their efforts in helping the student be successful, the chances that the student will succeed are greatly diminished.

2. Empathize with the teacher. We have all had difficult teaching experiences: failed lessons, problematic student behavior, lack of resources, and exhaustion. Becoming a teacher who forms positive relationships with students and adapts their teaching based on their knowledge of a student is not an innate skill—it is something learned through effort. Just as it is important for mathematics teachers to model mistake-making as a necessary part of learning, we need to do the same professionally for our colleagues.

3. Admit ignorance. I use the word ignorance here, but not in a negative sense. Instead, we need to admit to others that we haven't experienced what they have. Although we may have similar schools, students, or backgrounds, what has worked for us in the past may not work for someone else. When I began talking to the first-year teacher at the workshop, I was excited to learn about his experiences being immersed in that particular tribal community. My experience with Indigenous students generally had been limited and I was eager to hear his unique perspective. Even if I had lived and worked within an Indigenous community, my perspective would still be limited: Literature often treats Native American students as one large group, or even extends the generalization to Indigenous people globally, but we need to recognize that tribal identities vary over time and space (Masta, 2018).

4. Avoid essentialism. It is human nature to make generalizations and seek patterns. The harm comes, in part, when we assume that people will conform to our preconceived notions. Yet it is a common pitfall in culturally responsive teaching to rely on generalizations, commonalities, and past experiences. Educators looking for a helpful list of educational strategies for African American, Latinx, or Indigenous students may find such lists, and even attempt to integrate specific strategies into their teaching repertoire. However, that is not being culturally responsive. In reviewing the research literature on culturally responsive schooling (CRS) for Indigenous youth, Castagno and Brayboy (2008) caution that:

> ...although the plethora of writing on CRS that we review here is insightful, it has had little impact on what teachers do because it is too easily reduced to essentializations, meaningless generalizations, or trivial anecdotes – none of which result in systemic, institutional, or lasting changes to schools serving Indigenous youth. (p. 942)

Just as deficit thinking can often be identified when educators make absolute statements about groups of people, well-meaning essentialisms can be identified by blanket statements like "Indigenous students don't talk directly to adults; it's part of their culture." In *The Silent Indian as a Cultural Production*, Douglas Foley (1996) cautions against replacing "the old assimilationist narratives with anti-assimilation narratives that glorify [Native Americans'] heroic political resistance," effectively replacing "one simplistic cultural image with another" (pp. 88–89).

5. Offer research-based strategies. A general, research-backed approach to preparing teachers for diversity in the classroom—in particular, when they are a outsiders to that diversity—is a three-step approach. First, teachers need to examine their own beliefs and attitudes about teaching and learning, as well as their own cultural background, learning experiences, and privilege (McIntosh, 2003; Oskineegish, 2015; Sleeter, 2017; Villegas & Lucas, 2002). Second, learn about and engage with the community of your students. What makes this community unique? How can you connect with your students' culture to create a third space of possibilities, where mathematics meets your students' lived experiences? (Gonzalez et al., 2001; Moje et al., 2004). Finally, maintain an attitude of openness, flexibility, humor, and patience (Oskineegish, 2015).

These, of course, are general guidelines. For specifics, share the readings, podcasts, novels, movies, and songs that have inspired, educated, and challenged you. Direct colleagues to organizations that work toward reform in mathematics, like TODOS: Mathematics for All, Alliance of Indigenous Math Circles (AIMC), and the more familiar National Council of Teachers of Mathematics (NCTM). Provide them with the position statements of these organizations; these statements serve as a terrific introduction to the thinking and rationale underlying mathematics education equity work (Association of State Supervisors of Mathematics et al., 2021; National Council of Teachers of Mathematics, 2000; TODOS, 2016).

6. Continue the conversation. I have often considered how I could have approached this critical conversation differently. There are any number of ways to engage productively with colleagues about equity issues, and an equal number of unproductive ways. My biggest regret is not offering this teacher my email address. I sensed that this teacher was professionally isolated, out of ideas, and in need of a mentor. Instead of exchanging contact information, I walked away. I don't even remember his name, where he was from, or what he cared about. Unfortunately, that door has been closed.

Resources

Emily shed light on an unforeseen conversation that took place during an equity-focused professional development workshop with a colleague who expressed negative generalizations about his students. She emphasizes the significance of such moments as opportunities to engage in dialogues with peers, challenge harmful assumptions, and cultivate a deeper understanding of their beliefs and instructional practices. Table 7 provides resources to help with how teachers talk and think about their students.

Table 7. *Suggested Resources for Challenging Harmful Assumptions*

RESOURCE TYPE	RESOURCE	SUMMARY	PURPOSE
Video	*Build a School in the Cloud* by Sugata Mitra https://www.ted.com/talks/sugata_mitra_build_a_school_in_the_cloud?	A video in which Sugata Mitra shares his research with "Hole in the Wall" computers. He found that children create "self-organized" systems of learning when provided with a computer and internet access. This learning happens whether the children can understand any English (the default language of the computer when installed). It even happens when children have never seen computers, let alone been taught to use them.	To demonstrate how to assume positive intentions. When faced with challenging student behavior, whether it is a withdrawn student or one that is "misbehaving," many teachers conclude their students don't care about learning. Mitra reminds us that children are naturally curious and able to learn, despite any number of barriers to education. Mitra touches on many different aspects of formal education, but perhaps most importantly, his work reminds us that learning can be joyous.

(*continued*)

Table 7. *Suggested Resources for Challenging Harmful Assumptions* (continued)

RESOURCE TYPE	RESOURCE	SUMMARY	PURPOSE
Journal article	*The co-construction of opposition in a low-track mathematics Classroom* by Victoria Hand	An article that analyzes the co-construction of an oppositional culture in an 8th-grade, low-level math classroom over the course of one school year. Through an ethnographic analysis of classroom observations, videotaped lessons, interviews, and conversations, the author shows how opposition and defiance between teachers and students developed over the course of the year.	To demonstrate how to empathize with the teacher. Interpretations of verbal exchanges provide insight into how our actions as teachers may have a quite different effect than what we intended.
Essay	White privilege: Unpacking the invisible knapsack by Peggy McIntosh. In *Understanding prejudice and discrimination*	An essay on white privilege which lists 50 "unearned entitlements" that are conferred on white people in the United States and other cultures with long histories of white power dominance. It explains why people who benefit from unearned entitlements often fail to recognize them.	To demonstrate how to admit ignorance by inviting teachers to interrogate the history of racism embedded in our educational systems.
Video	*The danger of a single story* by Chimamanda Ngozi Adichie https://www.ted.com/talks/chimamanda_ngozi_adichie_the_danger_of_a_single_story	A TED Talk in which Chimamanda Ngozi Adichie describes the misunderstandings that arise between people when they adopt stereotypes as reality. She offers her own perspectives from both sides of the single story: the person making assumptions about another group and the person whose story is rewritten.	To demonstrate how to avoid essentialism. No matter what our single story about another group of people is, "it makes our recognition of our equal humanity difficult." To be effective teachers, we have to see the humanity and diversity in our students.
Professional organizations	*TODOS: Mathematics for All, Alliance of Indigenous Math Circles* (AIMC), and *National Council of Teachers of Mathematics* (NCTM)	These three math education organizations are committed to proactively working to make mathematics accessible to all students.	To offer research-based strategies. Professional organizations act as curators of academic educational research for their members; they also provide a wealth of strategies to use these ideas in the classroom.

REFLECTING FORWARD

Applying the Microlevel Injustice Framework

Use these guiding questions for your independent thinking and group conversations to move your thinking and actions forward around disrupting injustice.

<table>
<tr><td>Section 1:
Analyzing a Narrative</td><td>1. Where did Emily experience conflict, internal and/or external?
2. When could Emily have chosen to disengage with her colleague?</td></tr>
<tr><td>Section 2:
Weigh the Balance</td><td>3. What prevents you from addressing similar moments? What negative consequences could result?
4. How many students could experience positive change in their mathematics education by convincing just one teacher of a perspective shift?
5. Read the research section of the chapter that connects to the larger issues in education. What is at stake in choosing not to engage or respond?</td></tr>
<tr><td>Section 3:
Consider Your Context</td><td>Consider how your context may differ from Emily's.
<table>
<tr><th></th><th>Emily</th><th>You</th></tr>
<tr><td>Race</td><td>white</td><td></td></tr>
<tr><td>Setting</td><td>Off-site professional development workshop</td><td></td></tr>
<tr><td>Social justice background</td><td>Social justice focus began in high school</td><td></td></tr>
<tr><td>Years of teaching</td><td>12</td><td></td></tr>
</table>
6. How might the differences between your and Emily's context change how you would respond to a fellow math educator who has a deficit perspective?
7. What assets can you uniquely bring to the situation? (E.g., personality, knowledge of the community, class, education, etc.)</td></tr>
<tr><td>Section 4:
Authentically Disrupt Injustice</td><td>8. Which recommendations for action resonate most with you?
9. Which are most challenging? What steps will you need to take to overcome the perceived challenges?
10. What resources provided in the chapter could build new learning that will help you?</td></tr>
</table>

EVAN M. TAYLOR

Prelude

In 2021, Tyler the Creator released his Grammy nominated mixtape, Call Me If You Get Lost with DJ Drama, and it was centered in Tyler's analysis of his development as an artist in the music industry. While many songs caused me to pause and reflect on my own development as a teacher, the one that sat with me the most was the song entitled MANIFESTO. In the track, Tyler responds a bit to fans and outside observers wanting him to conform to certain archetypes that have been laid out for Black artists, specifically musical artists. In his lyrics, he reflects on the fact that although he cannot connect with the monolithic trope of Black masculinity, especially in hip hop, he still experiences and navigates the world and industry as a Black man. This sat with me because as I entered the teaching profession, and through the upcoming narrative, I often wore identities I thought I needed to be perceived as a social justice mathematics educator.

I am a Black male educator with eight years of teaching experience. I would wear Black Lives Matter shirts; I would teach lessons that I pulled from Rethinking Mathematics (Gutstein & Peterson, 2006), and I would read books and articles and listen to podcasts to prove to others that I was doing the work. In the development of my pedagogical stance, I centered the experiences of Black boys and men and continue to grow in learning how to develop spaces with Black girls and their experiences in mind. This year, a student had me stop, listen, and ask myself if I was aligned with my values and all the things that I professed and posted. I had read and been reading books on teaching and learning with Black children, specifically Black girls, and convinced myself that I had reached a point where I could begin to model for others, only to find that I had (and have) more work to do.

Narrative

It was a typical day of mathematics instruction with my first section of 6th-grade math learners. Students entered the room 5 minutes late because of their late release. When they finally arrived at the classroom, they took out their math notebooks, began to work on the first problem of the day, and prepared themselves to take notes on the material to be covered. As they began to write, the conversations started to dwindle and transform from conversations about what transpired in the last class to what was needed to proceed through the current mathematical problem. Some students wrote expeditiously so that they could have the opportunity to respond to the questions as soon as they were asked. Some students wrote moderately so that once those who had responded to the questions had given their answers, they would be ready to write their responses down in their notebooks. Some students sat and waited for the activity to end so that they could log onto their computers to play some games. Then there was me, ready at the side to get started. As soon as I spoke, waves of hands extended into the air to answer my questions. I was already racing against the clock because I felt the need to cover the instructional materials within the shortened time block. There was not really much space for them to intervene or influence the conversation because I needed to get to the exit ticket, which required students to answer questions so they could apply what they learned for the day before our 40 minutes expired.

Once I asked the first question about the warmup problem, we were off to the races, and everyone had to keep pace. Those that were willing to match the tempo raised their hands and raced their pencils across their math notebooks. Those who already knew that the tempo was not made or catered to them simply sat back and let other students engage in the conversation. I had convinced myself from the beginning of the year that I was embodying equity because I was pulling from various resources to develop a more responsive curricular experience, but I was still racing against the clock. I felt that to prove my worth in this new school space, I would need to get test scores that would prove my own value and the value of the students. This was all that was on my mind when I would teach and develop lessons. I was so focused on pressing ahead that I did not always consider how inequitably the classroom was operating. While I was continuing to push the pace of the lesson, a student raised her hand and asked, "Can you slow down a bit?"

This was a challenging moment because my district had called for acceleration instead of remediation due to the pandemic. The district mandated that teachers cover one lesson per day and that any student who missed concepts catch up through engaging with online platforms at home. Her question and request that I slow down a bit was asking me to put her needs and the needs of her peers at the center of instruction.

Parents had already emailed me numerous times asking that I provide time for students to complete the exit tickets in class so that they would not forget to complete them. That was in response to my opening exit tickets for 12-hour windows so that I could give feedback to students and use the data to make decisions for the next day's lesson. Before my student even finished her question, my mind was already set on the fact that we did not have time and that her question was preventing us (me) from reaching our (my) goal. Parents wanted me to keep the pace so that their children could master the standards, but at the same time slow down to give their children grace as we transitioned back to in-person learning.

I responded to her by saying that I could slow down a bit, but the notes would be available online after school and that she and others could get the notes down in their notebook later. I told her that if she could not get all the notes down, then that was not too concerning, and she should focus more on engaging in the dialogue and less on getting down every symbol that I wrote in the notes. I thought that this was an equitable response and would enable me to continue at the pace I had set in previous class sessions. It was the response that I had emailed to parents earlier when they said their children were having difficulty keeping up with the pace of the class. This was the response that I believed was aligned with the district's demands and a few of the parents' preferences to both speed up and slow down. I had convinced myself that I had done right by her and that we could move on in the lesson. In all honesty, I was not really listening at this point and was trying to offer the quickest solution and response to the situation so that I could still give them time at the end to complete the exit ticket. I was listening to her, but at the same time calculating which questions and conversations I would have to skip to complete the lesson.

Since I thought the issue was resolved, I was shocked when she stopped me again. She then responded, "That is not it. If I do not get a chance to answer [or engage] with it during class, I will not be able to understand what is happening." That moment was frustrating for me because it was causing me to really sit with myself and ask myself who I thought I was. Before this moment I was convinced that I was a critical mathematics educator who taught in ways that enabled students to be their full selves and to be able to engage in mathematical discussions. I read books on critical mathematics education and mathematics identity and agency and was sure that I was the math teacher that I had envisioned in my own mind. I had gone to conferences and talked to other math educators about how to develop and sustain communities of practice with students, but here I was being called out by a student to account for my inequitable actions. This was not the first time a student, specifically a Black girl, challenged me and implored me to change my teaching practices.

Earlier in my teaching journey, I donned the mask of Black male educator that many wanted me to wear. I was a no-nonsense type of educator who prided myself on how I could wield my voice to demand a room. I learned to pride myself on the amount of work and tasks I would require students to complete and the effort I would put in to have them maintain an elevated level of excellence measured through production and performance. One summer while engaging in these practices, I taught a Black girl who was not responding to the way that I was teaching and interacting, and I was convinced that I would have to make her stay after so that I could persuade her that the way that I was challenging and pushing the students was what she needed. In the dialogue, she ended up telling me that the way I was teaching and trying to motivate her did not work, and that I would have to find alternative ways to motivate and engage her in the mathematics classroom. After that conversation, I shifted a bit in my pedagogy and my practices, which showed me that I was still learning.

In the moment, I did not throw everything out and change my teaching style, but now she caused me to question myself and to ask who I am as a mathematics educator and as a person in the community who works with youth. It made me wonder and notice how inequitable situations can cause me to lean into versions of myself that I thought I had grown past. At the same time, it had me question how I make a classroom space equitable for those that may need varying paces and why it is we were in a race. In previous lessons and years of teaching, I had done this very same thing and taught at the very same pace. Amid teaching this way in previous years, I was traveling to various conferences and talking to other teachers about engaging in work that I was not even consistently engaging in myself. I told myself that if I set inequitable conditions to get through some material, then I could balance out my inequity by having a social justice lesson or conversation.

I entered the classroom with this idea that I was a certain type of educator and I often refused to really look into the mirror of my students' eyes to see who and how I really was. It was easy to go around saying the right things and quoting certain authors, but it is much more difficult to live in those values consistently. I thought that this challenging work of developing equitable learning spaces was more outer work, but I am constantly reminded that it is mostly (if not solely) inner work, and that the work is messy.

Reflections

The next day, I sat down and composed an email to her and her parents to address what had happened in the room and to speak about how their daughter caused me to pause, reflect, and adjust my mathematics teaching practices. Looking back now, it is bittersweet because a good portion of my mathematics teaching practices have come from many Black girls that I have taught. This is the most recent situation, but I have had many conversations in the past where Black girls have let me know what I need to do to support them in reaching their mathematical goals and in building their mathematical identities.

Morrison (1992) writes of how American literature had been positioned and imagined such that one of its main purposes was to center and preserve white male views. Mathematics learning spaces have historically been developed to position and privilege the stories, experiences, and identities of white men (particularly those who have European ancestry) (Battey & Leyva, 2016). In order to develop a gaze that decenters the mathematical identities and experiences of white men solely, we as educators must include the herstories of Black girls and women into our curriculum.

Through my teaching experiences, I learned that it is important to center past, present, and future Black girls and women in mathematics spaces, which can happen not only through reading but through engaging in conversations with Black girls about their experiences in mathematics learning spaces (Nazemi, 2022). Scholars who have engaged in interviews and dialogues with Black girls have found ways in which educators should best work with Black girls in teaching and learning with them (Collins et al., 2020; Joseph et al., 2019). Developing a Black girl gaze is a process in which one learns to listen to Black girls and women about their experiences and begins to analyze how one may be perpetuating systems or practices that may be causing harm to the Black girls. I am learning the complexities of developing this gaze in a classroom full of multiple identities and experiences, which is why I used the terms "towards". Kumashiro (2000) writes:

> We are not trying to move to a better place; rather we are just trying to move...Although we do not want to be (the same), we also do not want to be better; rather we want to become constantly, we want difference, change, newness. (p. 46)

I have grown beyond believing that I will ever fully develop a Black girl gaze, but I am determined to constantly become more aware of better ways to teach and learn beside Black girls and women.

As a Black male teacher and learner of mathematics, much of my teaching practice is rooted in my experiences and identity as a Black male, which differs from the multiplicity of experiences and identities of Black math learners, specifically Black girls. This is one aspect of my teaching that I never really thought about as a student teacher or even in my first years of teaching because I was extremely invested in the narrative that Black children needed me and other Black men to save them. That narrative impacted my relationships with each and every student that I taught as well as my ability to be an equitable educator and develop learning communities.

In the Netflix documentary *Jeen-Yuhs* (Simmons & Ozah, 2022), which follows the early career of Kanye West, there is a moment when Donda West, Kanye's mother, speaks to Kanye and cautions him that when the giant looks in a mirror, the giant sees nothing. Donda was warning Kanye of his growing ego and getting caught up in his own reflection. I fell for what Donda warned Kanye about and got caught up in my own reflection and ego of being a Black male educator so much that I could not see or hear the voices of my students. I could not hear the voices of the Black girls in my class because I was so caught up in perpetuating my own image and identity within the Black boys that I was teaching.

As Villenas (1996) reflected on her dualistic identity as both a colonizer and the colonized, she wrote:

> As some enter the ranks of teachers, administrators, and scholars, we are becoming the enforcers and legitimators as well as the creators of official knowledge...We scholars/activists of color need to understand the ways in which we manipulate our multiple...identities and how our identities are manipulated. (p. 728)

As a Black male educator, the stories that I believed in my early teaching career convinced me that I was solely the colonized, that I could not be the colonizer, and therefore could not be causing harm because I occupied a Black body. My experiences in teaching and learning from Black girls invited me to navigate, explore, and develop a more complex and critical look at myself and my own inconsistencies.

Developing a Black Girl Gaze

I cannot teach in new ways until I critically reflect on my previous learning experiences in school spaces and make a commitment to dismantle and disrupt those practices throughout my teaching practices. Halberstam (2013) writes:

> ...when you seek to make things better you are not just doing it for the Other, you must also be doing it for yourself. While men may think they are being 'sensitive' by turning to feminism, while white people may think they are being right in opposing racism, no one will really be able to embrace the mission of tearing 'this shit down' until they realize that the structures, they oppose are not only bad for some of us, but they are also bad for all of us. (p. 10)

In developing a Black girl gaze, I engage in reflexivity, which is challenging when I aim to look and reflect on myself and my own practices honestly. As I even write these sentences and reconstruct my reflections, I still must combat the need to develop and present a clean-cut story that places me at some point of arrival in which readers will be comforted by my journey. In my representations and reflections on how Black girls came to view me, I am challenged to write their words through my own eyes but also reflect on my own positionality within my reflection.

Villenas (1996) notes how she could not escape her "experiences of marginalization and dislocation...at the same time [she] cannot escape the privilege afforded to [her] as a university professor (pp. 75–76). As I compose this piece as a Black male teacher, I cannot escape my experiences as a Black man and former Black boy in mathematics learning spaces, but at the same time, I cannot escape the privilege afforded to me as a mathematics teacher. As a Black boy in mathematics learning spaces, I had to navigate the challenges of my Blackness while still enjoying the privileges of masculinity in STEM spaces not afforded to Black girls. As I teach and reflect on my experiences as a mathematics educator, especially as a Black male educator, I have the privilege to develop classroom spaces that develop and foster the mathematics identities of Black boys. The challenge, however, is that in doing so I may neglect developing spaces conducive to the development of spaces that center Black girl experiences as well. King (2003), in writing about stories, tells of a Nigerian storyteller who said:

> In a fractured age, when cynicism is God, here is a heresy: we live by stories, we also live in them. One way or another we are living the stories planted in us early or along the way, or we are also living the stories we planted--knowingly or unknowingly--in ourselves. We live stories that either give our lives meaning or negate it with meaninglessness. If we change the stories we live by, quite possibly we change our lives. (p. 153)

The way I changed the stories I lived by was by seeking out stories that did not coincide with my developing worldview and view of myself. Carson (2013), in her novel *Red Doc*, writes that "to live past your myth is a perilous thing," and I am finding this to be true as I reflect and am beckoned to reflect on the myths that I have believed and taught through.

Resources

Evan reflects on his intersectional identity as a Black cisgender man, persistently confronting patriarchy while supporting young Black girls in mathematics classrooms. Table 8 has resources that provide teachers ways to incorporate instructional approaches that support Black girls in mathematics classrooms.

Table 8. *Suggested Resources for Supporting Black Girls in Mathematics*

RESOURCE TYPE	RESOURCE	SUMMARY	PURPOSE
Documentary and panel discussion	*Picture a Scientist* Full Panel Discussion	A documentary and panel discussion that offers teachers a view into the experiences of women in STEM fields and how they navigate STEM spaces to open opportunities to others to explore various careers in STEM.	To help educators develop an understanding of how students' gendered and racialized identities impact how they navigate spaces, and to encourage them to look for mentors to speak with their students.
Conference presentation	Get on up from the page by Masie L. Gholson	A conference presentation on how to support Black girls in mathematics classrooms situated in research from the North American Chapter of the International Group for the Psychology of Mathematics Education Conference in 2019.	To enable teachers to analyze ways in which Black girls experience math learning environments.
Article	*How Black Girls Benefit When Math has Social Interaction and Ways to Learn Together* by Kara Newhouse	An article that highlights work from scholars who engage in research on how to support Black girls.	To educate mathematics teachers about what Black girls need from them, which includes cultivating a humanizing mathematics classroom.
Article	*I'm scared to ask for help: Black Girls' Voicing of their K-12 Math Experiences* by Ashli-Ann Douglas, Mariah Harmon, Nicole Joseph, and Bethany Rittle-Johnson	An article that shares Black girls' perspectives of their experiences in the mathematics classroom.	To remind teachers of the importance of listening to their Black girls in the classroom.
Journal article	*Clean corners and algebra: A critical examination of the constructed invisibility of black girls and women in mathematics* by Maisie L. Gholson	An article that discusses the racialized and gendered experiences that Black girls face in mathematics classrooms.	To provide a robust understanding of the concerns that Black girls have when learning mathematics.

REFLECTING FORWARD

Applying the Microlevel Injustice Framework

Use these guiding questions for your independent thinking and group conversations to move your thinking and actions forward around disrupting injustice.

<table>
<tr><td>Section 1:
Analyzing a Narrative</td><td>1. What aspects of the narrative resonated with you?
2. What fears did Evan have to consider when rethinking his teaching practice?</td></tr>
<tr><td>Section 2:
Weigh the Balance</td><td>3. Read the research section of the chapter that connects to the larger issues in education. How is the unjust structure you oppose not only bad for students from underrepresented backgrounds in mathematics education but also bad for you?</td></tr>
<tr><td>Section 3:
Consider Your Context</td><td>4. Consider how your context may differ from Evan's.
<table>
<tr><th></th><th>Evan</th><th>You</th></tr>
<tr><td>Race</td><td>Black</td><td></td></tr>
<tr><td>Gender</td><td>Male</td><td></td></tr>
<tr><td>District pressure</td><td>Acceleration</td><td></td></tr>
<tr><td>Tenure</td><td>New to school</td><td></td></tr>
<tr><td>Years of teaching</td><td>8</td><td></td></tr>
</table>
5. Have you encountered a similar tension in your classroom that started with a student request for an instructional or curricular change?
6. What aspects of Evan's background or identity are different from your own and would affect how you choose to respond?</td></tr>
<tr><td>Section 4:
Authentically Disrupt Injustice</td><td>7. Responding to injustice in mathematics education often requires teachers to be vulnerable and brave. What are some possible risks that could be taken?
8. What next steps can you take to listen to your students' needs?
9. What resources provided in the chapter could build new learning that will help you?</td></tr>
</table>

CHAPTER 7

This is White Folks Mathematics: What Do You See?

DR. KENDRICK SAVAGE AND DR. TASHANA HOWSE

As a new high school teacher, I was assigned a classroom next to an experienced teacher, Ms. Johnson, a middle-aged white teacher who would become my mentor. I absolutely appreciated Ms. Johnson as she had taken me under her wing. It wouldn't take long to develop a respect for her given the atmosphere of the school. She showed great care for the students in a predominately Black school and wanted what was best for them. I saw firsthand how hard she worked and the mathematical activities and risks she took in the classroom to get students engaged and motivated about mathematics.

One afternoon, the bell rang to change classes, and Ms. Johnson stepped outside of her classroom. We engaged in a discussion about a scenario between her and an African American male student in her math class. She shared that she noticed Malcom was not doing his math work, so she went over to help him, and he said to her, "This is white folks mathematics." She had responded, "Well, what about Mr. Savage?" But the student replied, "Oh, that's different."

Caught In Between

This short conversation immediately triggered an internal struggle within me. Although I listened with an intent to respond with something of substance, I couldn't help but hesitate because of the thoughts I was wresting with in that very moment. I understood where the student was coming from, but I didn't want to say anything that I thought would fracture my relationship with Ms. Johnson. Being a Black man with a similar background as the student, I knew what he meant and how he must have felt. But how could I address such an issue with someone who has looked out for me? Questions ran through my mind. Could I ask her if she thought about her role in why a Black student might say the comment? Would this create a Black versus white moment? Would she look at me differently? Would she lose respect for me? I was quickly losing my grip on being able to speak openly and honestly.

Connecting My Education History

About 6 months prior to this very conversation, I was a graduate student working on my PhD in mathematics while attending a predominantly white institution. This was a big deal for me and my family. Neither my mother nor my father held high school diplomas. Privately, I started carrying this weight that I had to make it to help my family, so completing my journey of earning my PhD in mathematics was everything to me. I wanted my mother, who never graduated, to see her son walk across the stage to receive his PhD in mathematics. I wanted my parents to feel proud that I did something with their sacrifices, and I was excited to start a graduate program that I knew I would finish.

However, it wasn't long into my PhD program before I started facing unexpected adversity—adversity that I couldn't even explain. I was accused of cheating on a homework assignment. I went to the professor, disputed the accusation, and he bluntly told me whose paper I had copied. She was a white girl. It was so outrageous that I couldn't believe it. I stood my ground as firmly as I could. Eventually, he changed my score after I showed proof that I hadn't cheated, but still refused to give me full credit.

Ultimately, I was not able to complete my PhD in mathematics at that school, but not because I didn't want to. I went through so much trying to do well in the program and at the same time survive the unfairness. After I began to argue about the unfairness of how I was being treated, the mathematics chair pulled my record and told me I had not performed to the program's standards. I received a phone call telling me that a committee of mathematics professors voted that I was to finish the semester but not return after it. I was beyond hurt. My hopes and dreams were shattered! Every time I lay down for the next 2 to 3 weeks, I cried in silence. Being asked to leave my doctoral program in mathematics created a lot of emotional pain.

Processing My Internal Struggle

My experience with my doctoral program allowed me to resonate with Malcom's comment in Ms. Johnson's class, because he likely also felt that Black males were not welcomed in math spaces. Recently confronting my white professors also left me less confident in speaking up. However, I understood Ms. Johnson's perspective as a seasoned teacher and a mentor I valued.

Mostly, I was a bit shook as to what my influence at the school had been. I was a young Black man in mathematics, teaching at a predominantly Black high school. I had assumed that my presence alone was making a difference, to the point where students wouldn't think that mathematics was just for white people. I was conflicted. Why did the student say this? I'm right here! Every day I solve problems, carry myself with respect, and return the next day to make an impact in spite of what I have gone through. So, am I making an impact or not? These are the questions that left me feeling confused as Malcom's words played over and over again in my mind.

Prior to this moment, I knew I had made a difference. I had students every year that would switch into my class, but I never had a student who left my class to go elsewhere. There are multiple factors that I believed played into this. I was the only Black male mathematics teacher. At the time, I was young and understood the culture of the students and for the most part looked similar to their age. Students in the school knew I was very strong in my mathematical content knowledge and that I was extremely patient and motivating with the students. My own upbringing and family background helped me to know what they were up against, and I shared about challenges openly with them. My students understood that I cared.

After my first year, all the new teachers that year were faced with losing their jobs due to district cuts. During this time, so many students came to me to tell me how sorry they were that I was going through that. I even ended up in the newspaper because a student mentioned my name as a teacher who would be impacted by this situation. Many parents contacted me to share their concern of me possibly losing my job. To my knowledge, I was doing something right. I was making a difference. The students could see that, or so I thought.

How I Responded

After listening and going through my own personal turmoil, I responded with the safest answer I could think of—"Really?"—and ended the conversation. Now, as I reflect on this moment, I can't help but feel like I let that particular student down as well as all the other Black male students that Ms. Johnson would teach. I should have asked what the context of the problem was, what she did to relate the problem to student experiences, and what she knew about the student's personal life. Then, I would have become more of an ally by assisting with cultural understanding. I realized the issue was not about her ability to teach, but possibly her ability to culturally connect. Specifically, culturally connecting the mathematics with the student experiences.

My mentor teacher, responding with, "Well, what about Mr. Savage?" unknowingly dismissed the student's Blackness altogether because it still didn't give any opening into his specific experiences or how he couldn't relate to the problem. By my answering the question so safely, it did nothing to improve the situation. This meant that if there were other students feeling this way in her classroom, then she would receive the same results. I had failed to improve a situation that was in my power to help improve.

At the time, it made sense for me to answer the question as I did. My graduate school situation put me through a lot. I never wanted to feel that way again. If I had answered the question by saying that she didn't recognize that's how they see math, I could have possibly created friction with my mentor, and I didn't want that.

Overall, a moment of impact such as this one is not always easy to navigate. But I think conversations like this are definitely worth having and can be done with no issues at all. My experiences and what I had gone through got the best of my thoughts in the moment. It's understandable why I responded the way I did. It didn't make me wrong. It made me human. It didn't make me right. It caused me to grow. Reflecting on the moment, I can give the advice to let respect lead the way and do what is best for the students. I also believe there is always a right time to correct or improve upon that which can be improved, meaning that whether you decide to have additional conversations later that day, the next day, or later in the week, you must have them. African American men have gone unseen for far too long, and this is still evident today.

Organizations, such as Black Lives Matter, are constantly having to remind the world that Black lives matter. But why? Why is this necessary? Isn't this clear to all? These questions were answered May 25, 2020, in Minneapolis, MN, when the world witnessed the brutal killing of George Perry Floyd, Jr. The answer is no! The death of African American men like George Floyd and Ahmaud Arbery ushers in an interesting connection between Black men inside and outside the classroom. An interesting paradigm is created because African American boys in the mathematics classroom could ask the same question—Don't Black lives matter?

RESEARCHER RESPONSE **DR. TASHANA HOWSE**

My research expertise was shaped by several educational factors. My educational experiences, for the most part, were all shaped while attending predominantly white institutions. My high school, for example, was also majority white. Therefore, from high school to college, there was no big difference in classroom experiences. Early on, specifically in high school, I never thought about the intersection of culture and learning, but I was very aware of the classes I enjoyed versus those I didn't. The teachers that appealed to me most were teachers that took time to get to know me. Additionally, I was intrigued by how they integrated home and personal experiences into the classroom content.

Characterizing Kendrick and Malcolm's Culture

Culture to me is more about character than race. A person's character is often connected to how they were reared, what they like and do not like, if they are an introvert or extrovert, and even what they find interesting. Put simply, there are multiple aspects of one's character that could give you an idea about the culture that surrounds them. It is not wise to assume anything about anyone and automatically think you know their culture. One cannot look at a Black man and say, "Oh, I know he will like this because Black people like this." They have no idea of the person's experiences, upbringing, or interest. Understanding someone's culture implies that one must first seek to understand and learn more about them. Yes, culture can be specific to certain groups, but that's when one finds a group of people that share similar experiences. Thus, if one wants to understand someone's culture, then one should get to know them before assimilating them into a certain group.

I want to express my appreciation of Kendrick for capturing such a critical moment in the classroom that deserves to be analyzed and expounded upon to bring awareness and knowledge to future educators.

He does an excellent job of situating the reader within the tug-of-war between his prior experiences and what he was feeling in that critical moment. Reading his words made me ask myself questions such as: "Why after all these years is this critical moment still so fresh in his heart and mind?" and "Does his connection with the student extend beyond both being Black men?"

It is worth mentioning that Kendrick is a first-generation college graduate. After having his college experiences and becoming a researching professor, he understands the total extent of what it means to be first generation. From the lens of a first-generation college student and now professor, he understands more about why each educational moment is so critical with first-generation college students. As with Malcom, the student in this commentary, first-generation students could be one negative moment away from giving up. Kendrick has shared multiple details regarding the struggles of matriculating through his undergraduate and graduate experience at predominantly white institutions. Therefore, this understanding of Kendrick's background makes it clearer as to why this issue continues to tug at his heart after all these years.

Unpacking Invisibleness in America's Classrooms

First-generation college students, defined by Hsiao (1992) as being the first in the family to attend college, are often met with challenges that family and non-first-generation college students rarely understand or experience. For instance, they are 4 times as likely to leave higher education without a degree compared to those who are not first generation (Engle & Tinto, 2008), and are often met with difficulties before they leave home as they battle with "the culture of home and the culture of higher education" (Hsiao, 1992).

As a Black man who attended predominantly white institutions and then taught at a predominantly Black school, it was refreshing to hear him express his vulnerability of positioning himself in Malcolm's shoes. He knew exactly how Malcom was feeling and why Malcom expressed those words. Black students, especially boys, have been overlooked and ignored for far too long, particularly in cases where their culture is absent in the content. Tate (2006) highlights this perfectly as he captures the words of Carter G. Woodson:

> And even in the certitude of science or mathematics it has been unfortunate that the approach to the Negro has been borrowed from a "foreign" method. For example, the teaching of arithmetic in the fifth grade in a backward county in Mississippi should mean one thing in the Negro school and a decidedly different thing in the white school. (p. 31)

Woodson's words highlight the feeling of so many African American students who have sat in classrooms and have gone unrecognized, especially African American men. This feeling of being unrecognized is what we call feeling invisible. Invisibleness, in education, specifically mathematics education, does not have the potential to promote interest or student engagement. If anything, it perpetuates student disengagement and lack of motivation to pursue the learning of mathematics. Although never intentional, being overlooked in the classroom still happens more than it should.

Instructional Practices That Combat Invisibleness

Savage (2020) found that the most important factor for student mathematical motivation in predominantly African American high schools was the teacher. Therefore, teachers must be as well established in effective teaching practices as possible. Teachers cannot hide behind not connecting because they are a different race with different experiences. Though that may be true, teachers must work to make up the gap, or at the very least give a considerable amount of effort toward cultural growth.

Researchers have studied instructional practice and teacher effectiveness for years. Haycock (1998) pointed out that:

> If we but took the simple step of assuring that poor and minority children had teachers of the same quality as other children, about half of the achievement gap would disappear. If we went further and assigned our best teachers to the students who most need them there's persuasive evidence to suggest that we could entirely close the gap....

Greer and Mukhopadhyay (2003) stated that, "There is more and more concern about the gap between school mathematics and the lived experience of students and the adults that they become" (p. 2). Mathematics teaching that operates with the absence of student experiences only creates less opportunity for students to advance in society. Mickelson, et. al., (2013) stated that:

> Mastery of mathematics by a nation's youth is essential for individual and societal advancement. The generation and use of innovations in the fields of science, technology, engineering, transportation, medicine, public health and safety, and commerce require citizens with mathematical knowledge and skills. This nation's ability to prepare the next generation to fully participate in civil society and to enter the increasingly technical workplace requires the public-school system, where the vast majority of U.S. children receive their formal educations, to successfully teach mathematics to all children.

Teachers that are not prepared to teach mathematical content with the student in mind may offer a one-size fits all approach. Mathematics presented in such a way is considered a foreign approach. Shared from a mother's perspective, Martin (2009) states:

> This system is not designed for African American children to learn. This system is designed for European children and if you were to design an Afrocentric method of teaching these children, they would excel in math, you know, they would do better. And I believe that until the day I die. That is the reason why these kids are not doing well....African Americans have got to stop giving so much power to this system; that's why we are in the condition that we are in. (p. 327)

Berry (2003) acknowledges what happens when we don't try to reach Black students with a foreign approach, "When used properly, knowledge about African cultural style provides educators with a means for interpreting students' thoughts, feelings, and actions, while raising expectations for student success." (p.246). Berry goes on to explain that:

> African American learners use concrete imagery as a means of reckoning reality. They draw on their daily experiences to facilitate learning. African American learners classify ideas, items, and experiences based on how things relate to each other. In addition, African American learnings prefer experimentation, improvisation, and harmonious interaction with others environment...A cultural-based perspective to teaching mathematics recognizes that within the African American culture there is an affinity for rhythm and pattern...not all cultural expressions should be mathematized, but it is important to help students see the connections between mathematics and their contextual and cultural ways of knowing. (p. 247)

Dr. Martin Luther King Jr. (1967), very aware of what relinquishing your power meant, expressed some powerful words:

> Many Negroes have given up faith in the white majority because "white power" with total control has left them empty-handed. So, in reality the call for Black power is a reaction to the failure of white power. It is no accident that the birth of this slogan in the civil rights movement took place in Mississippi—the state symbolizing the most blatant abuse of white power.

African American students have long worn the cloak of invisibleness inside of classrooms. Traditionally, much of that has been passed down through an educational system that has neglected to situate the identity of African Americans within what is taught and how curriculums should be developed. The words by Dr. King remind us that history will repeat itself if we don't push for equity for all, and the importance of being seen must be placed on the entire human race, not just the majority.

Equitable Teaching Practices Promote Visibleness

First, Kendrick does a great job of naming this critical moment by introducing the term *invisibleness*. He was able to see what his white teacher colleague could not see, which was how students can be present, yet invisible all at the same time. In his pre-civil rights novel, *Invisible Man*, Ellison (1952) highlights the feeling of what invisibleness means for African Americans by stating:

> I am invisible, understand, simply because people refuse to see me...when they approach me they see only my surroundings, themselves, or figments of their imagination-indeed, everything and anything except me...that invisibility to which I refer occurs because of a peculiar disposition of the eyes of those with whom I come in contact...you often doubt if you really exist. You wonder whether you aren't simply a phantom in other people's minds...And, let me confess, you feel that way most of the time. You ache with the need to convince yourself that you do exist in the real world, that you're a part of all the sound and anguish, and you strike out with your fists, you curse and you swear to make them recognize you. And, alas, it's seldom successful. (pp. 3–4)

The eloquent words of Ralph Ellison capture this moment perfectly. When Malcom states that this is "white folks mathematics," he is really striking out with his fists, cursing and swearing, just to get his teacher to recognize what is happening. Ms. Johnson, responding with "What about Mr. Savage?" only perpetuates Ellison's words "when they approach me they see only my surroundings, themselves, or figments of their imagination-indeed, everything and anything except me...." (Ellison, 1952, p. 3). Again, she never saw Malcom nor what his words truly meant.

Sadly, this moment happens all too often. Maybe for the next teacher it is not about "white folks mathematics," but about believing that Black students have more supportive programs or more scholarships designed for them. Whatever the case may be, there are educators in positions that are responding with answers grounded in ignorance or misunderstandings that are affecting students. James Baldwin, a Civil Rights Movement leader and author, once said:

> There is a very bitter prescience in what this boy-this "bad nigger"- is saying, and he was not born knowing it. We taught it to him in seventeen years. He is draft age now, and if he were not in jail, would very probably be on his way to Southeast Asia. Many of his contemporaries are there, and the American Government and the American press are extremely proud of them. They are dying there like flies; they are dying in the streets of all our Harlems far more hideously than flies. A member of my family said to me when we learned of the bombing of the four little girls in the Birmingham

> Sunday school, "Well, they don't need us for work no more. Where are they building the gas ovens?" Many Negroes feel this; there is no way not to feel it. (p. 77)

Though Baldwin's words describe a different era, it is his words of "We taught it to him in seventeen years" that are so eye-opening. African Americans have been assimilating to jobs, education, and rules over the years that were never in their best interest. The fact that Malcom uses the phrase "This is white folks mathematics" is him standing up to and against this constant teaching that leaves him feeling invisible! He decides at this moment that he will not assimilate to powers that exclude his presence and consequently ignore his power. Do you think Malcom simply came to that conclusion only in that moment, or had he been taught this from grade to grade? Baldwin's words are powerful and full of truth—the curriculum and teaching of African Americans has been so alienating that one starts to wonder if they are really needed.

Kendrick, who highlighted his own educational difficulties in graduate school, was in fact used to not feeling needed. He, himself, had felt the heat of the gas oven, the pressure to be perfect, not to rattle any feathers, unless he wanted trouble that could come in the form of broken relationships, people taking away his opportunity, or simply not liking him. It would have been great if in that critical moment, Kendrick could have explained the truth of that moment to his colleague, if he could have found the words to educate his colleague with grace instead of feeling the weight of what speaking his mind would cause. Then, perhaps in that moment, he would have kept another African American student from striking out with their fists, cursing and swearing, just to be seen.

Resources

Kendrick and Tashana examine how an African American boy's statement in his mathematics class not only revealed his invisibleness, but also revealed his white teacher's inability to address his racialized comment. The resources in Table 9 provide insight on engaging *all* learners by bridging learners through their lens and connecting it to who they are. By highlighting the importance of relationships, these resources support educators in establishing and maintaining equitable teaching practices that promote individuality and visibleness, i.e., culturally responsive teaching.

Table 9. *Suggested Resources for Establishing Equitable Instruction*

RESOURCE TYPE	RESOURCE	SUMMARY	PURPOSE
Book	*Choosing to see: A framework for equity in the mathematics classroom* by Kyndall Brown and Pamela Seda	A book that provides a framework for creating equitable mathematics experiences for all students.	To give a practical framework to teachers who want to engage in equitable instructional practices to engage students.
Book	*For white Folks Who Teach in the Hood... and the Rest of Y'all Too: Reality Pedagogy and Urban Education (Race, Education, and Democracy)* by Christopher Emdin	A book that is about a Black man who experienced invisibleness in urban classrooms and felt undervalued by his teachers.	To support educators in understanding the importance of relationships with students through the use of culturally responsive teaching in mathematics classrooms.
Book	*The Equity & Social Justice Education 50: Critical Questions for Improving Opportunities and Outcomes for Black Students* by Baruti K. Kafele	A book that engages readers in deep reflection on critical questions related to the teaching and learning of Black students.	To gain an understanding and appreciation for having an equity mindset and teaching for social justice, especially when teaching Black students, as well as how to use it to empower students of color.
Book	*Engaging in Culturally Relevant Math Tasks: Fostering Hope in the Secondary Classroom* by Lou E. Matthews, Shelly M. Jones, and Yolanda A. Parker	A book that provides practical lessons ideas for engaging students in culturally relevant mathematics tasks which are connected to students' lived experiences.	To empower teachers to draw upon knowledge of their students to design equitable mathematics learning experiences.
Book	*The Brilliance of Black Children in Mathematics: Beyond the Numbers and Toward a New Discourse* edited by Jacqueline Leonard and Danny B. Martin	A book that offers a historical perspective of the connection between Black children, their lived experiences, and mathematics.	To show teachers the importance of knowing their students and capitalizing on that knowledge to drive instructional decisions.
Book	*High School Mathematics Lessons to Explore, Understand, and Respond to Social Injustice* by Robert Q. Berry III, Basil Manley Conway IV, Brian R. Lawler, and John W. Staley	A book with well-designed lessons that promotes social justice education. Each lesson walks students through how to use mathematics to solve problems as well as issues related to injustice in their local communities.	To gain an appreciation for culturally responsive teaching and why it is important for students to see themselves in mathematics.
Webinar	*Supporting Culturally Responsive Pedagogy with IM K–5 Math™* by LaToya Byrd https://illustrativemathematics.org/im-k-5-math-as-a-support-for-culturally-responsive-pedagogy/	A webinar that explains how an illustrative mathematics resource is designed with culturally responsive pedagogy in mind and capitalizes on the assets of students to drive and move learning forward.	To learn about specific components that support teachers in implementing culturally responsive mathematics in their classroom.

(*continued*)

Table 9. *Suggested Resources for Establishing Equitable Instruction (continued)*

RESOURCE TYPE	RESOURCE	SUMMARY	PURPOSE
Webinar	*Why Teach Mathematics Through a Culturally Responsive Teaching Lens?* by Sunil Singh https://www.nationalacademies.org/event/11-02-2021/why-teach-mathematics-through-a-culturally-responsive-teaching-lens	A webinar that expresses the importance of teaching mathematics through a culturally responsive teaching lens.	To empower teachers to capitalize on the cultural experiences of their students.

REFLECTING FORWARD

Applying the Microlevel Injustice Framework

Use these guiding questions for your independent thinking and group conversations to move your thinking and actions forward around disrupting injustice.

Section 1: Analyzing a Narrative	1. Why did Kendrick experience internal conflict from a short conversation?
Section 2: Weigh the Balance	2. Read the research section of the chapter that connects to the larger issues in education. Are there students in your classroom who might feel invisible?
Section 3: Consider Your Context	Consider how your context may differ from Kendrick's. **Race** — Kendrick: Black; You: **Gender** — Kendrick: Male; You: **College history** — Kendrick: First-generation college graduate; You: **Relationship with colleague** — Kendrick: Mentor who has been helpful; You: **Education** — Kendrick: Started a doctorate program in mathematics; You: 3. Can you recall a moment when you wish you had spoken up but chose not to? What were the reasons that prevented you from doing so?
Section 4: Authentically Disrupt Injustice	4. Responding to injustice in mathematics education often requires teachers to be vulnerable and brave. But for many educators of color like Kendrick, simply being present in a math space is a challenge in and of itself. Are there moments to see yourself with compassion even if you chose not to respond to an injustice? Explain your response. 5. How can you respond to a similar situation that can challenge a coworker to learn while also honoring your own needs to feel emotionally and physically safe?

Section 3 table:

	Kendrick	You
Race	Black	
Gender	Male	
College history	First-generation college graduate	
Relationship with colleague	Mentor who has been helpful	
Education	Started a doctorate program in mathematics	

CHAPTER 8

Challenging the (a*)Political Mathematics Classroom

MICHELLE LO AND DR. TERESA DUNLEAVY

...When we are discussing a human rights issue and how to help, and [s]omebody follows up with support towards a person who believes systemic racism doesn't exist, [who] threatens the safety of BLM protesters, [who] calls protesters thugs, and [who] does nothing to help when [responding] to racially charged murders...I really don't think that's a cool thing to say.

—Peter, Personal Communication

This quote was written in an email from one of my students, Peter, to me, Michelle Lo. In this chapter, I discuss what happened when I reminded my precalculus students to vote in the 2020 presidential election, and one of my students responded by writing "#Trump2020" in the Zoom chat. I battled whether to address this comment in the moment and afterwards. Below, I highlight how Peter's email demonstrated a voice and agency that led me to facilitate a restorative circle about the harm that was caused by the political statement.

We use "*a**" in our chapter title to emphasize and stand for "always." While mathematics is often perceived as political, we adopt the "always" perspective of Gutiérrez (2013) to suggest that math is *always* political. That is to say that, as humans, we reveal our political choices about what counts as mathematics through the ways that we enact the teaching, learning, and practice of mathematics as a content.

The Incident

On August 23, 2020, Jacob Blake, unfortunately another Black man, was shot by police officers in Kenosha, Wisconsin and left paralyzed from the waist down. I first thought of my Black students. I thought about how I would address that harm to my classes the following day; it felt like uncharted territory for me as a beginning teacher.

On the one hand, I felt hesitant about speaking up. I identify as a Taiwanese American daughter of immigrants. In my household, voting was rarely discussed—whether it was who to vote for, whether to vote, or what to vote on. My parents instructed my siblings and I to keep our heads down and to avoid conflict, something that I continue to struggle with today, particularly as I think about speaking up about injustices. Further, what place did discussing racism have in the mathematics classroom? What did my students, their families, and my administration think about discussing racism in a mathematics class?

On the other hand, I knew, as a teacher teaching for justice, I could not stay silent, especially after the deaths of Breonna Taylor, George Floyd, and Ahmaud Arbery earlier that year. I could not stay silent because of the anger I felt at our country—staying silent meant that the violence toward Black lives was normalized to the point where it was not even acknowledged. I could not stay silent because alongside the anger sat an ache for the Black community. I felt a type of ache that signaled what was at the core of humanity: empathy and compassion; nobody should experience this level of violence. In the end, I decided to be intentional about speaking up. I decided I would not simply acknowledge Jacob Blake's situation only to acknowledge this injustice as a one-time incident; instead, I felt that it was important to address the systemic roots that led up to this police shooting.

On the following day, I acknowledged Jacob Blake in my classes, tying this tragedy to other recent incidents of police gun violence. I told my students that policies surrounding guns needed to change. I then remembered that most of the students were either nearing or already eligible to vote in the upcoming presidential election. "Remember to vote in this upcoming election," I said. "You all have so much power to change the way things are." One student shared resources in the Zoom chat on registering to vote. Then, a white male student, whom I will call Aaron, typed "#Trump2020" into the chat. I froze. What should I say? How should I respond? In a split-second moment of panic, I moved on. I let the moment pass, knowing I and the rest of the class had seen what Aaron had typed. Worse, the class knew that I chose not to respond.

Class and School Community

At the time of the incident, I was in my first year of teaching. It was also the 2020-2021 academic school year, when the majority of the school year took place virtually due to the COVID-19 global pandemic. I had just completed a master's and teacher credentialing program focused on social justice and equity. I taught at a public high school in California I will call Beachside (all names are pseudonyms). During that year, Beachside enrolled about 1300 students; school profile data reported that approximately 60% of the students were Hispanic, 20% were white, and about 5% of students were Asian, African American, and multiracial. Beachside was a Title I school where about 40% of the students were considered socioeconomically disadvantaged, although local reports indicated the surrounding area was increasingly gentrifying. I also completed my student teaching at Beachside.

Although the incident occurred during the third week of the school year, this precalculus class had already developed a particularly strong level of engagement and participation. On the first day of school, I realized that several students already knew each other. And not unlike other classes I taught on Zoom, most students' cameras were turned off during class, even though most students had met and knew each other prior to the pandemic. The daily check-in included a slide saying: "How are you doing on a scale of 1-5? Drop it in the chat." Students engaged in the chat both privately and publicly, asking questions, sharing ideas, and occasionally turning on their mics to speak. Throughout the first

2 weeks of school, I also worked with this class to cocreate a list of "Class Commitments" (Lo, in press). The general culture of this particular precalculus class was high-energy, talkative, and interactive.

As I progressed from student teacher to teacher at Beachside, I noticed increased efforts among our faculty to learn about and work towards educational equity. After the murder of George Floyd on May 25th, 2020, in Minneapolis, Minnesota, and alongside the continuation of the Black Lives Matter movement through that summer and fall, Beachside established student and staff equity teams. Beachside set a school goal to work to address anti-bias and create an anti-racist campus and community. Other goals included initiating conversations about race among and within staff and students. And while student teaching in the 2019–2020 school year, administrators started to lean away from out-of-school suspensions and toward restorative discussions with students. I also found that the administration and teaching staff collectively supported and believed in the school's mission, as demonstrated in part through strong support of staff-wide professional development on anti-racism and Courageous Conversations. Courageous Conversations refers to a protocol developed by Glenn E. Singleton and colleagues in which facilitators support community members to engage in meaningful racial dialogue (Singleton & Linton, 2007).

Alongside the increased attention to racial justice, the school community also continued to experience instances of injustice and conflict. In the first 3 weeks of school, prior to the incident, some students reported bullying and discrimination on their social media accounts, and others shared anonymous confessions. The confession posts ranged from romantic crushes to racist and homophobic comments about other students. Although it was unclear to me whether students faced any school consequences, the fall 2020 Beachside administration sent out a video asking the community to come together as a Beachside family. In previous years, homophobic comments turned into physical fights, which resulted in student-written letters to district leadership.

Processing the Incident

Perhaps partially due to the work that had been happening at the school level, a sense of guilt washed over me after that class ended. I once again battled over whether to speak up, and if so, how to respond to Aaron's comment. What did it mean to speak up about Jacob Blake but then to be silent about a student's support for then President Trump? Should I respond during the next class? If I did follow up, how would Aaron feel? Part of my battle was worrying that Aaron would feel politically and socially alienated, which could lead to disengagement with the community and in turn, in the class. I also worried about whether students' family members might be concerned about me pushing a political agenda. Lastly, I thought about how the Beachside administration might react to my suggestion that students vote. Would I be jeopardizing my job? Growing up, my friends and I would try to guess our teachers' political affiliations; we believed that it was frowned upon for teachers to share any political views and that for many teachers, it was generally easier to not talk politics. As a beginning teacher, I still believed that I could avoid conflict by not talking about politics, although Beachside's mission of working towards anti-racism and anti-bias granted me some degree of safety in bringing up voting.

Simultaneous to addressing the incident, I also felt the pressure to continue with the week's precalculus lessons. Beachside's virtual bell schedule also meant that we had less instructional time than during years when classes met in person. I felt that I had opened a can of worms by reminding students to vote. While I did not regret bringing up voting, I realized that not following up would also, in and of itself, be a political statement. I decided that if I did not follow up, I would be committing an injustice to my students, including Aaron, both as learners in our community and as humans in an already unjust society. Not only would I be perpetuating this injustice through silence, but I would also be dehumanizing the students whose identities had been attacked by former President Trump and by others during the era of Trumpism, including my students of color, students who are undocumented, and students who identify with the LGBTQ+ community.

Recalling that I had chosen not to be silent about Jacob Blake, I realized I would be politically and morally at odds if I chose not to follow up about the implications of #Trump2020. Furthermore, mathematics shapes how many students view themselves, including how they engage with their identities as learners. For many students, experience with mathematics shapes their identities in ways that are marginalizing and traumatizing (Aguirre, et al., 2013). Ultimately, I decided I could not and would not ignore Aaron's comment.

My decision was further confirmed by Peter, the student whose quote opened this chapter. A larger excerpt from Peter's email reads:

> ...I really don't want to come off as a petty student. But, I feel like I should let you know as a part of your learning community. As always, thank you for standing up for what's right and being not only a great teacher but a great person. I always feel welcome in your class. However, there's one teeny problem.
>
> I'm pretty certain you saw Aaron's comment in the chat saying "Trump 2020." I'm not the type of person to immediately dismiss a person based off of their political life, but, when we are discussing a human rights issue and how to help, and [s]omebody follows up with support towards a person who believes systemic racism doesn't exist, [who] threatens the safety of BLM protesters, [who] calls protesters thugs, and [who] does nothing to help when [responding] to racially charged murders...I really don't think that's a cool thing to say....We have a lot of POC and LGBTQ+ students in our class and I think it's important for you to even just have that in your mind somewhere. (Peter, personal communication, September, 2020)

Initially, I felt defensive and alarmed that Peter had emailed me. But after rereading his email multiple times, I grew more grateful and amazed at the voice, power, and agency that Peter exhibited to call out this injustice. Not only did his email highlight some ways the incident made students from marginalized groups feel, but it also underscored what I saw as my obligation to work toward creating a safer space for the students of color and queer students in my classes. And while I was already agonizing about what to do next, Peter's email helped me find the strength to address the incident in the following class.

At first, I was at a loss for ideas on how to follow up. When I was growing up, mathematics was simply about abstract numbers and solving for unknown variables. I attended public K–12 schools in Oregon in a suburban neighborhood. My mathematics classes were void of conversations around social justice, politics, and how mathematics can be used to minimize inequalities. This trend continued in my undergraduate mathematics classes until I started taking coursework in education.

Throughout graduate school and now in my own professional development, I have thought about how to teach mathematics for social justice (e.g., Gutstein, 2005; Kokka, 2015) as a means to play the game in order to change the game (Gutiérrez, 2016). As a beginning teacher and forever learner, it has been helpful for me to find communities of teachers with similar social identities and teaching philosophies, including the #miseducAsian and #EduColor spaces on Twitter, as well as the Nepantla Teachers Community. These spaces have allowed me to connect, read, and hear about other educators' experiences in teaching for social justice.

Throughout my time as a student and now teacher, the incident was the first time I called out and publicly discussed an overt injustice. I no longer saw mathematics as apolitical. I realized, in part because of the incident, that my experiences growing up contrasted sharply with my new experiences learning in community with other educators. I could not stay silent about injustices in mathematics learning spaces.

I knew I had to do right by my students. Given my limited experience responding to injustices in the classroom, I decided to call my mentor teacher from the previous year to ask for advice. The advice that resonated with me the most was to "be authentic and show up in the way that you can." I used that advice to adapt a practice I had learned during my master's program: restorative circles (Center for Restorative Process, 2015)." In this practice, community members come together to address harmful behavior through a process of building community.

Responding to the Incident

Beachside operated on a block schedule, so the next class occurred 3 calendar days after the incident. I started class by sharing a slide to initiate the restorative circle (see Figure 1).

Figure 1. *Class Slide After the Incident*

We are a community

1. Well this is awkward (for me mostly)
2. We are here to learn from each other in a safe space.
   a. Impact > Intent
3. I'm sorry
4. Moving forward ... what's (not) allowed?
   a. Basically anything hurtful towards one's identity: racism, homophobia/transphobia, xenophobia, ableism, etc.
5. I'm here for you.
   a. Office hours are a great time to speak to me individually
6. Any other thoughts?

With the majority of students present, I reminded students of the incident while leaving out Aaron's name. I decided to use the slide in Figure 1 to show up as my authentic self while balancing the tension between how students with marginalized identities might have felt alongside Aaron's sense of belonging. I thought about how I started class by taking off my own armor as a way to demonstrate that this was a conversation in which I allowed and welcomed vulnerability. I shared how I felt about my silence and my position as a teacher in the room, I apologized for not saying anything in the moment, and I reminded the class that good intent is not enough. I then shared what was and was not allowed in my classroom, stating that while political differences were allowed, discrimination and implications of discrimination were not. I shared how important it was to be conscientious of our words and to respect one another. Lastly, I shared that I would make time for individual conversation and reflection. I talked for about 10 minutes, during which time there were no comments in the chat and no students came off mute.

While a typical restorative circle might use a talking piece (Center for Restorative Process, 2015), when I opened up the floor to the rest of the class, I asked students to use the raise hand feature on Zoom. Although students left their cameras off, students started offering their ideas both in the chat and by speaking aloud. One white student mentioned that the incident was an inappropriate response to Black Lives Matter and a police shooting because it implied further support for racism. Another white student next said something like, "I'm sure it was just a joke; people don't need to be offended so easily."

Aaron then wrote into the chat that he did not mean to hurt anyone intentionally, and that he just simply supported President Donald Trump. A fourth white student said that political views matter but should not be brought into the classroom since everyone was just trying to learn, and that by sharing political views, it could be triggering or harmful to some students. Several students of color then chimed in to agree with the fourth student's sentiment, some using the heart emoji, and some writing something into the chat like, "^^." Following the string of agreements that sharing political views could be retriggering or harmful, another student, John, who identified as Hispanic, reiterated the need to create a safe space for learning while acknowledging that everyone has different political opinions. After about half an hour, the conversation died down. I thanked everyone for sharing. At that point, Aaron left the Zoom class and did not return to class that day.

Ms. Lo's Reflections

In reflecting now, I acknowledge my own positionality as a woman of color amongst a classroom of majority students of color, as well as the status that I held as the teacher in this space. At the same time, I believe this restorative circle was far from perfect. While I set the table to welcome vulnerability and hoped to include all voices, a majority of the voices dominating the discussion were white students, especially those who chose to unmute their mics. I was also particularly mindful that the entire conversation occurred virtually. As I have reflected further, I have thought about setting up a virtual circle in future opportunities so everyone could visually see students participating (Yu, 2021). A virtual circle would also offer the opportunity for everyone to speak or pass (as in a traditional restorative circle) without having to take initiative to speak out. If I had held the restorative circle in person, I would have asked students to sit in a circle and use a physical item as a talking piece, drawing on the power of physically sitting in a circle in community (Center for Restorative Process, 2015).

I also did not anticipate Aaron would leave class after the discussion. While I never confirmed with Aaron why he left, I also did not follow up with him individually to reflect on either the incident or the class conversation. If I could go back, I would invite him to talk about how to help him feel more included in our learning community.

I continue to think about whether it would have been best to respond immediately to the incident. Specifically, I think about whether responding live to the incident, instead of the next class day, would have made the situation better or worse. The *Toolkit for Interrupting Oppression* (Oregon Coalition Against Domestic & Sexual Violence, 2018) offers five different ways to interrupt oppression: immediate action, actions over time, getting/providing support, indirect action, and no action. Considering my teaching experience and positionality, I have reflected that I did not feel ready to respond the first day in an authentic and intentional manner. Having time to reflect, consider my values, come up with a structure for reflection and restoration, and resume after having some time to process directed me toward action. I have also reflected that each critical moment might render different advantages and disadvantages to ways of interrupting oppression.

Throughout the rest of that school year, I continued to wonder about how students felt regarding the classroom community. The class as a whole continued to have lively and energetic engagement, and Aaron continued to put an effort into classwork. I still wondered how students felt when placed into a breakout room with Aaron, and how I would navigate whole-class conversations. I also thought a lot about my interaction with Peter. I decided to follow up with Peter a year later, during the 2021-2022 school year, hoping to learn whether the restorative circle affected the class environment. Below is an excerpt from our conversation in which Peter discussed how the restorative circle helped him learn, in part, about another classmate:

Ms. Lo: How do you think having that [restorative circle] conversation affected the classroom environment?

Peter: Yeah, I think it just, like, let us all know what is tolerable and what isn't? And also, that kind of having that conversation kind of shows each other everyone else's colors...John's brother is actually like a huge Trump supporter. I'm not sure if you're aware... But anyways, [John] stood out to me because he would, he spoke out against kind of - Aaron's behavior. If not, if it weren't for that, I wouldn't have known the kind of contrast that is between John and his brother. So, in that way, it kind of made it feel a little safer. I do know John was, like, very into our class. He was like, he was like a big person in our class.

Ms. Lo: Mhmm. Do you think it helped you get to know your classmates in a different way? And also, because maybe there's some shared beliefs that it felt more inclusive or safe?

Peter: Yes, definitely...that's exactly what I'm trying to say. We all saw that we sort of shared the same beliefs. And it was, it felt like a safe space. (Peter, personal communication, December, 2021)

And while Peter talked about how John stood out to him because John called out Aaron, I was simultaneously surprised when several students of Color seemed to agree to not bring politics into the classroom. I interpreted this to mean that students did not think that teachers or students should explicitly state who they were supporting in upcoming elections. At the same time, my conversation with Peter revealed that having political conversations in mathematics class was critical in creating a classroom environment where students felt seen beyond mathematics content, devoid of context. While I believe that stating one's political beliefs is not always necessary, I do believe that classrooms are inherently political. As educators, I believe that it is our duty to create classrooms that are actively inclusive of students from all types of backgrounds, particularly students from underrepresented backgrounds. This means that as a teacher, it is necessary to not only advocate for norms and values that include all students, but also to enact the norms and values that include all students in the mathematics classroom. Beyond inclusivity, teachers' and students' decisions about who and what has power and status in the learning of mathematics also demonstrate how classrooms are political spaces.

For Peter, the restorative circle helped reveal that he was not alone in the belief that implications of racism, homophobia, and other forms of identity-based discrimination should not be tolerated in the classroom. The restorative circle attended to building community through responding to harm; it also reinforced ways to listen to and share perspectives. In his email, Peter also mentioned the significance of discussing Aaron explicitly because "as a community, we really do need to kind of…like each other to an extent, because it is a collaborative environment. So, we do, in a sense, need to not hate each other" (Peter, personal communication, September 2020). As a classroom community, I wonder how we could have moved beyond tolerating one another and toward disrupting injustices. I continue to wonder: How does a classroom community move towards collaborating and learning together not despite our differences, but because of our differences?

RESEARCHER RESPONSE **DR. TERESA DUNLEAVY**

I, Teresa Dunleavy, am a white woman, a professor of education, and a former high school mathematics teacher. I faced many difficulties throughout my childhood and believe it was the connection I found with learning that led me toward a career in education. I also acknowledge that, in spite of challenges from my personal life, I benefit from white privilege and systemic racism on a daily basis. I started my career wanting all students to be successful, but I quickly noticed differing patterns of achievement for students, particularly as they related to their racial and ethnic identities, along with their varied access to economic resources.

My frustration in not being able to significantly change my students' experiences led me toward my current work. Initially, my research focused on classroom practices that strive toward equity (Dunleavy, 2015). I know all students can do high-level mathematics, but their different ways of being brilliant are not always fostered in traditional environments. I have grown my work toward understanding the systemic oppressions that lend themselves to a narrow view of smartness, competence, and achievement in mathematics (Dunleavy, in press). I center my work on an anti-racist agenda that offers counternarratives for dismantling dominant patterns of oppression and investigating how white supremacy reinforces dominant perspectives of who can do and who has access to mathematics (Dunleavy, in press).

Ms. Lo initially applied to a master's program of which I was codirector. Although she completed her master's elsewhere, we stayed in touch, maintaining similar interests in teaching, research, and justice in education. We now share a research project investigating the role of complex instruction in either perpetuating or dismantling patterns of white supremacy in mathematics teaching (Lo & Dunleavy, 2020). Complex instruction is a cooperative learning pedagogy developed by Elizabeth Cohen and Rachel Lotan focused on equitable classroom practices that position every learner as highly capable (see Cohen & Lotan, 1997; 2014).

Math is Always Political

I often bring the question "Is math political?" to students in my education courses. While many mathematics education students know and love mathematics, some are tied to dominant ideologies that position mathematics as beautiful but devoid of social context. The ability for a person to separate politics from the content of mathematics is grounded in embracing dominant ideologies as normal. Gutiérrez's (2013) discussion of the sociopolitical turn frames the tensions around how knowledge, identity, and power dynamics occur in classrooms, towards an acknowledgement that mathematics is *always* political. To echo our definition of political, it is important to acknowledge *what* mathematics we learn with *how* we conduct mathematics teaching and learning—each choice is placed within the political realities of our contexts.

For Whom Do We Teach?

Mathematics teachers are faced with constant decisions about who they are responsible to teach toward (Chen et al., 2021). When racist and oppressive comments are made in our classes, as teachers, we face opportunities to be silent or to act. And while Ms. Lo was drawn to call out the incident as oppressive, she wrestled with her own feelings about encouraging students to vote in order to change oppressive laws, alongside worrying if her job was at stake because she had made political comments. At the same time, as her reflection continued, including when she received the email from

Peter, she moved forward, eventually deciding that attempting a restorative circle was a critical opportunity to disrupt the injustice that some students felt they experienced in the incident. During the next class, Ms. Lo decided to show up for her students as herself—both there to teach Aaron and to support all the students who felt the injustice of Aaron making a statement about President Trump immediately after discussing the police shooting of Jacob Blake.

Ms. Lo worried about isolating Aaron, saying, "Part of my battle was worrying that Aaron would feel politically and socially alienated." Often educators wrestle with questions about whether their job is to serve all students. And in particular, this often means that we hold in tension whether and how to support our students who have political views that differ from our own, and with whether oppressive comments should be minimized to different political views.

Restorative practices (e.g., Center for Restorative Process, 2015; Kokka, 2019) can support students and teachers to face oppressive classroom interactions and can move communities toward new connections. In Ms. Lo's restorative circle, she first centered her own humanity, acknowledging and apologizing for the silence that had occurred. Next, she named her intent—for students to learn from one another in ways that made them feel safe. In so doing, she acknowledged that the incident had made some students feel unsafe. By explicitly naming that statements that are harmful toward individuals' identities are not allowed, Ms. Lo acknowledged the inherently political nature of the learning environment while explicitly stating that oppressive comments would not be welcomed. She also offered an opportunity for students to meet with her individually. It is important that Ms. Lo felt supported by her school community to disrupt what she felt was an unjust comment made during her class, in part because she wondered about the security of her job and in part because attending to confrontation was counter to how Ms. Lo was raised.

Ms. Lo noted that she led the virtual restorative circle by allowing volunteers to share. She also noted that, at first, it was mostly white students who were talking. Although the students ended the circle's conversation by agreeing with not having political conversations in class, this might have been because students wanted to avoid conflict and not feel unsafe, as they had before. Jones and Okun (2001) discussed this as the fear of open conflict characteristic of white supremacy culture. In this case, instead of openly discussing the harm they felt by discussing the shooting of Jacob Blake and addressing, in their precalculus class, the pattern of police brutality against Black people in America, the conversation shifted toward creating a space that is kind. And while Peter's email to Ms. Lo focused on the systemically racist and oppressive experiences of students of Color and

LGBTQ+ students in the Beachside community, the class conversation during the restorative circle stayed grounded on students getting along, which Peter later cited in his follow-up interview with Ms. Lo as a need for a class that involves interaction to get along, saying, "We really do need to kind of...like each other" (Peter, personal communication, December, 2021).

As a scholar, I note that the process of disrupting injustices is layered; a teacher's ability to engage pedagogical responsibility during complex sociopolitical moments is shaped by their personal histories. Teachers who continue to work toward making sense of what they teach alongside for whom they are teaching may have the ability to disrupt patterns of injustice in sustainable ways.

Resources

Ms. Lo's story describes how student voice and agency called forth a restorative circle about the harm inflicted by a student's political statement towards queer students and students of color during an online class session. This instance challenges the notion that mathematics and teaching are apolitical spaces and highlights the need to make room for courageous conversations. Table 10 provides resources that help teachers disrupt oppressive comments.

Table 10. *Suggested Resources for Disrupting Oppressive Comments*

RESOURCE TYPE	RESOURCE	SUMMARY	PURPOSE
Book	*Ratchetdemic: Reimagining Academic Success* by Christopher Emdin	A book that offers strategies for teachers to be both ratchet and academic.	To empower students and teachers to embrace their identities.
Book	*Rise Up! How You Can Join the Fight Against White Supremacy* by Crystal Marie Fleming	A book that discusses the origins of racial injustice alongside contemporary and institutional examples of white supremacy in action.	To raise awareness about the enduring impact of historical injustices on marginalized communities and how these legacies continue to shape society today.
Teacher Organization	Nepantla Teachers Community	A community of teachers whose mission is to use social justice in their classrooms.	To provide a sense of community. This group organizes annual cohorts centered on different themes (e.g., navigating racism, math classrooms as political, etc.) for the purpose of developing math teachers in social justice practices.

(*continued*)

Table 10. *Suggested Resources for Disrupting Oppressive Comments (continued)*

RESOURCE TYPE	RESOURCE	SUMMARY	PURPOSE
Toolkit	Toolkit for Interrupting Oppression by The Oregon Coalition Against Domestic & Sexual Violence Toolkit for Interrupting Oppression	A toolkit that not only outlines considerations and guiding principles, but also offers activities for skill-building, encourages cross-cultural communication and humility practices, and advocates for the regular practice of interruptions.	To offer specific actions and strategies for disrupting oppression in a variety of scenarios, including phrases teachers and students can use to speak up.
Website	White Supremacy Culture: Right to Comfort, Fear of (Open) Conflict, and Power Hoarding. https://www.whitesupremacyculture.info/comfort--fear-of-conflict.html	A website based on earlier work from Jones & Okun (2001). The website is built as a resource for communities to use together to understand (and dismantle) white supremacy culture.	To connect to white supremacy culture characteristics, the right to comfort, fear of (open) conflict, and power hoarding, which the authors discussed in this chapter.
Webinars	Learning for Justice: Speak Up at School https://www.learningforjustice.org/professional-development/webinars/speak-up-at-school-0 Discussing Race and Racism with Students https://www.learningforjustice.org/professional-development/webinars/lets-talk-discussing-race-racism-and-other-difficult-topics-0 Responding to Hate & Bias at School https://www.learningforjustice.org/professional-development/webinars/responding-to-hate-and-bias-at-school	The Learning for Justice platform includes webinars that offer activist strategies for teachers and students.	To provide strategies for teachers and students that help with identifying hurtful language and how to intervene, as well as discussing difficult topics and responding to hate and bias.

REFLECTING FORWARD

Applying the Microlevel Injustice Framework

Use these guiding questions for your independent thinking and group conversations to move your thinking and actions forward around disrupting injustice.

Section 1: Analyzing a Narrative	1. What do you identify as the main tension(s) Ms. Lo experienced during the incident?
Section 2: Weigh the Balance	2. What fears do you have when addressing recent political events in math class? 3. Read Teresa Dunleavy's research section of the chapter that connects to the larger issues in education. What is your reaction to the claim that math is always political? Who benefits from the belief that math is neutral or apolitical? Who suffers from this belief?
Section 3: Consider Your Context	Consider how your context may differ from Ms. Lo's. (see table below) 4. What actions did Ms. Lo take that you wouldn't do yourself? What actions do you think you would take that Ms. Lo didn't? 5. How do your identity/social markers and history inform actions that feel more or less comfortable to take?
Section 4: Authentically Disrupt Injustice	6. How can you prepare political conversations while considering your own comfort and risk tolerance, rather than simply avoiding them?

	Ms. Lo	**You**
Race	Taiwanese American	
Parents' political beliefs	Avoid political issues and potential conflict	
Education	Master's and professional development learning around math education and social justice	
Years of teaching	1	
School demographics	Racially diverse, Title I	

PART 3

CHAPTER 9

Disrupting Injustice Through Your Critical Moments

The previous chapters provided the building blocks to better discern critical moments in your own classroom and school. Perhaps these narratives have illuminated weak points or given space to think about an unaddressed moment in the classroom. Naturally, you've started to think about your own practice and how to identify critical moments to disrupt injustice. Here, we lay the groundwork for forming and sharing your stories with other math educators. Each of us has potential critical moments and can benefit greatly from sharing about them in a community. We've taken the 4-section guide and repurposed each section to illuminate your critical moments for yourself and others.

Section 1: Identify Your Critical Moment

One way to begin is by thinking of an interaction during this semester or school year in which you were aware of an injustice. Sometimes this can be a difficult task, so it may be easier to access moments that elicited tension. It could have been a moment that you consciously chose to ignore for lack of better options or one that is a recurring issue. One of the challenges of identifying these potential critical moments is that they can seem innocuous or so commonplace that they go under the radar. Some of the examples provided in this book started with a small comment or observation.

There are four common types of interactions that may be able to help identify potential critical moments:

1. Teacher-Teacher: Teachers have many spaces where they interact with one another. Whether it's through a weekly grade-level meeting or in the teachers' lounge, teachers can propagate beliefs that suggest a specific population is inferior to others in mathematics. It may be through subtle statements about specific groups of students. Teacher-teacher interactions can also become critical moments in collaborative decisions, such as during an end-of-year meeting to finalize student recommendations for honors courses for the following school year.
2. Teacher-Supervisor: From time to time, teachers must also deal with their school administrator or supervisor when confronting injustice. This usually calls for skillful and politically minded conversations due to the power dynamic. Administrators may request or require a teacher to implement a policy,

curriculum, instructional move, or disciplinary procedure that is dehumanizing for students. Teachers can feel the additional weight of consequences when thinking about how to move forward shrewdly and tactfully.

3. Teacher-Student: Many critical moments stem from a student sharing a frustration or pain point about the classroom. Students raise issues ranging from grading fairness, pacing, favoritism, pedagogical styles, and curricular materials, among others. In these cases, teachers hold power in the relationship and can choose to ignore or engage with such students. Other moments can surface from a strong relationship a teacher has with a student, allowing the teacher to understand what students from underrepresented backgrounds may face more deeply.
4. Student-Student: Teachers also aid in student interactions with one another. Students might collectively vocalize views that demean a group as inferior or superior in mathematics and intelligence. A student can antagonize another student intentionally or unintentionally in math class. Teachers may find themselves in challenging and critical moments when faced with responding to injustice in student-student interactions.

Begin to detail the dialogue, reactions, and internal thoughts of the moment. As you form your narrative, consider what was challenging about the moment. There are many reasons that a critical moment is challenging. Maybe you felt pressured by the bell, or you didn't know what to say. Additionally, describe how you reacted in real time. This first reaction certainly doesn't have to be the only reaction. In several of the chapters, teachers decided to return to a moment and use it as a platform to have a deeper discussion.

Section 2: Weigh the Balance

Section 2 addresses the questions "Is it worth it?" and "How can I take hold of my power?" Note that disrupting injustice in critical moments is typically challenging because one is choosing to go against the grain. Math teachers need not feel ashamed to recognize that challenges and consequences exist. In continuing to build the critical hope referenced in Chapter 1, we take stock of our roadblocks honestly, prepare ourselves, and take calculated risks with integrity.

This is why it's so important to dissect the power dynamics in your critical moment. We can better determine the potential consequences, fears, and weak points. Each of the four types of interactions described in Section 1 situates the teacher in different places of power given their roles. Even aside from the role titles, there may be more nuanced power dynamics. Perhaps a fellow teacher is a mentor teacher or is the best friend of a supervisor. Perhaps a student's parent is on the local school board. Be prepared to share these kinds of details with your teacher community.

Aside from power dynamics that exist between individuals, there are also dominant societal norms to consider. For example, people don't think of mathematical spaces as places to talk about racism, sexism, or social influences. If a teacher chooses to bring up anti-racism in mathematics, there may be confusion about its relevance or pushback because it's not considered traditional content. Taking action against what has been normalized produces unpredictable results, which can be particularly straining for math teachers drawn to the field for its reliable and comforting logical processes. This isn't to say there's anything bad about such teachers, but that it will take more mental and emotional energy to commit to such actions, and that work shouldn't be minimized.

Whatever the case, we use this understanding of power to reveal the consequences we could face through a response to injustice. The easiest types of negative consequences to see are the tangible ones like being fired, having a meeting with an angry parent, or getting reprimanded. Further, some consequences are intangible and are so quickly processed that we may not even be conscious of them. Even though they are not as apparent, these intangible consequences are important to identify because they are used in our decision making. We name some of them here to make them more visible.

- Social: A teacher can be cast out by other teachers if they speak up against a popular belief of practice. They may get cold shoulders when they need someone to cover their class, borrow a stack of paper for copies, or access created lesson plans. These social consequences can extend to losing mentorship opportunities or fellow teachers to sit next to during all-staff meetings.
- Professional: Taking a stance against a lauded practice could tarnish a teacher's reputation. Others may not listen to their ideas or advice in meetings or leave them out of important decision-making initiatives.
- Teacher control: Teachers may choose to address injustice in classrooms but feel worried about not knowing how students will react. Students could respond emotionally, check out due to boredom, or instigate discord.

These consequences can be quite daunting, particularly when laid out at once. We do this to begin the process of problem solving. Once they are clearly labeled, they can be addressed. When we skip this section, we can prematurely conclude that a response is not worthwhile due to a vague sense that we are unequipped.

There are, of course, consequences in inaction as well. Injustices continue to perpetuate narratives that people are assigned competence and dignity based on their race, gender, or underrepresented backgrounds. And there is the consequence of not upholding our own integrity. When we are witnesses to injustice, we have an opportunity to put our values into action. Can we face ourselves if we choose to ignore it? With all these considerations, we decide on how to move forward with our humanity intact.

Section 3: Consider Your Context

When you share your critical moment with others, provide context. Here are some recommendations:

- Include the strengths and needs of your school community as it relates to your critical moment.
- Position yourself with a description of your own teaching background and education.

- Share how your race, gender, or identity as a whole may have affected the moment.
- Explain any particular influences in your formative years that shaped your processing of the event.
- Describe the assets you uniquely bring to the situation.

These details are what make your critical moment unique and provide the kindling for how you can disrupt injustice in your context.

Section 4: Authentically Disrupt Injustice

This section will remain purposefully nebulous because we, as authors, believe that as you go through Sections 1–3, you will begin to create unique responses and feel empowered to act on them. Analyzing the injustice and deeply interrogating our present context can lead teachers to respond in brilliant ways that have never been done before. This final section may be better to do in a trusting community of educators because disrupting injustice is a challenging and creative process.

As math teachers practice identifying critical moments and utilize these structures to learn through them in community, students will experience transformational mathematics because they are experiencing transformational people. We encourage you to use this guide in a regular cycle of reflection and action.

Claiming Your Moment

Changes in practice also pair with changes in heart. Usually at professional development workshops, there is at least one teacher who raises their hand (or quietly whispers to their colleague) and says, "This isn't going to work for my class." Afterwards, one group of teachers will explain why it won't work, but another group of teachers will contend as to why it can. The real thing that teachers are trying to say is "This isn't going to work for me." In this regard, they are right. Teachers must face themselves when making any kind of instructional change, and professional development will not be effective without addressing this notion.

The previous chapters have shared examples of what critical moments can look like and how we can respond to them. The real challenge comes from recognizing your own moments and leaning into them. Until then, the chapters in this book are only a mental exercise. It is serious work to reflect on the influences that have shaped our identities as mathematics educators and take ownership of our strengths or assets during critical moments. Disrupting injustice at the microlevel requires us to be present to ourselves and our students with authenticity.

If we are honest, we know that no one will hold us accountable to these microlevel injustices. We can ignore them, pretend we are unaware, or excuse ourselves from these moments. It's simply easier to go about our days through the path of least resistance. And yet, people enter the teaching field because they care about young people and want a better future for them. Teaching is not simply a job to clock in and out. It is a deeply human and relational occupation that gives us opportunities to care for one another.

You have the power to consider your critical moments of today. Perhaps there's a tension around the feedback you provide your students after an assessment, or perhaps you notice a student of color disengaged in the back of the classroom. Whatever the moment, you can begin the process of responding authentically without waiting for a policy reform or law amendment. We encourage you to hold space for your own moments, to share them with colleagues, and to begin a new kind of journey.

Conclusion

One of the takeaways from this book is that mathematics teachers can feel empowered to align their beliefs and actions toward social justice. Injustice and oppression are not only happening in the structures and policies of our society but also are occurring in daily interactions. As shown throughout this book, the contributors courageously shared their tensions, hesitations, and responses when attempting to disrupt inequities. They recognized that mathematics classrooms are places where complex interactions occur among various individuals. For example, critical moments can occur through a conversation between two teachers sharing a deficit perspective, between students in a comment about political candidates, or a student expressing they are not good at math because of their race.

As we conceptualized this book, we saw the benefit of talking with each other and the contributors. When teachers are asked to listen to other educators' stories, it's often in the context of success. It's much easier to paint a well-crafted picture in these snapshots of teacher life when a problem of practice is solved with an easy fix. But such solutions can feel frustrating for the teacher who hasn't found such success after expending great effort. Several of our editing remarks were through virtual meetings when we asked for vulnerable details about how they processed these challenging situations. During these conversations, instead of encountering composed educators with all the answers, we met people experiencing a range of emotions, such as uncertainty and even fear.

We noticed a pattern of the types of questions we asked our contributing authors. These questions were, "What consequences did you fear?" and "What factors (in your history, racial identity, community, etc.) contributed to how you responded?" In this internal excavation, we began to make sense of how these teachers made difficult decisions not as isolated examples of education superheroes, but as human beings leaning into their identity, history, and community.

One of the reasons these kinds of conversations and narratives are not examined is because teachers encounter a messy space where notions of right and wrong are not always clear. Even more unsettling is our claim that there's great value in listening to the stories of our peers that aren't replicable or scalable. At first glance, in Chapter 4 "Who's Good at Math?," it seems the instructional moves described have potential use for Asian American math teachers at a selective enrollment school of majority Black students. Although Esther had previous experience having professional conversations around racism, her experience is not transferable. Even in the case of accelerating a promising Black student like in Chapter 3 "A Gateway, Not a Gatekeeper," Tandrea was able to leverage her unique position as a teacher of both classes to ease the transition for Malik. These examples encourage teachers to move toward challenges thoughtfully and consider their positionality because disrupting injustice in schools is a local, contextualized process.

To be sure, anything that requires combined self-examination and action elicits a great deal of vulnerability. We must admit to ourselves our troublesome beliefs or actions. In Chapter 6's "Towards a Black Girl Gaze in Mathematics Learning Spaces," Evan saw himself as a model for equitable math teaching yet was concerned that his actions weren't lining up with his beliefs. He could have decided to zoom in on his successes and ignore the interactions that didn't integrate with his self-view. He chose, instead, to admit he didn't have all the answers, honor his students' perspectives, and redefine what a social justice mathematics educator looks like. Similarly, Michelle, in Chapter 8 "Challenging the (a*)Political Mathematics Classroom," initially felt defensive when her student emailed her about not speaking up during the last class. Yet she took a step back to recognize the perspectives of her students as valid and worth making room for. Math teachers, used to being known as competent and unbiased, must do the challenging work in examining their approach to ensure they are learning from their students, particularly when confronted by those from underrepresented backgrounds.

There are so many potential critical moments with our coworkers or fellow teachers that could alter their path or pedagogical approach along with potentially hundreds or even thousands of their students. We read in Emily's Chapter 5 "Face-to-Face Encounters: Reframing Deficit Thinking as an Opportunity for (Shared) Growth" how she chose to engage with a fellow teacher instead of walking away again. Capitalizing on these moments is not easy mentally or emotionally because we are making deliberate decisions to oppose a dominant narrative. These stories quietly run in the backgrounds of our minds with insidious phrases like "they would get it if they actually tried to learn English," "the Black kids always sit in the back to goof off," and "if their parents don't care, why should I?" Yet it's also in teachers' power to take the time to disrupt these unconscious beliefs or actions for our peers and even ourselves.

That's not to say that disrupting injustice lies solely on the backs of teachers. Teachers are under great pressure to perform, instruct, and converse by both explicit and implicit expectations. Kendrick's decision not to respond to a critical moment in Chapter 7 "This is White Folks Mathematics: What do You see?" was one he admits he would have changed. We share it here as a reminder that not every moment can have a response and that there are times to extend grace to ourselves.

Math teachers of color can feel isolated as they face unique challenges in their field. In my (Esther's) own experience facilitating conversations for math teachers of color, many have expressed a complex relationship with their race and mathematics. Black and Latinx teachers noticed that their peers do not think of them as highly competent or capable as mathematical thinkers. Moreover, many math teachers of color are met with more student needs. Like Kendrick, they are the only math teacher of color at their school, and many students of color look for their support as one of the few adults in the building with similar backgrounds or experiences related to their race . Sometimes simply existing as a math person of color in a majority white space is a disruption of injustice. This work of disrupting injustice is not limited to a single interaction, so we can make sense of the kinds of risks we are willing to take one critical moment at a time. These narratives remind teachers that disrupting inequity is more than relaying instructional ideas and changing pedagogical approaches. It is about promoting practices and interactions that uplift students from underrepresented backgrounds.

Education leaders promote solutions toward equity that are teacher-proof so students will learn the mathematical content regardless of who is in front of them. Likewise, some professional development is centered around grade-level curriculum, lesson pacing, or grading systems. While these reforms can be important, who is in front of students also matters a great deal. Math teaching is a primarily social activity that brings together many different people into the same space. These differences are what ultimately make our work creative. Teachers can begin to examine how they make ripples in their classrooms, schools, and lives, write their own story, and authentically show up with their unique assets to every encounter. The micro level interactions described in this book can greatly impact the ways teachers engage with students and impact much needed macrolevel change. With enough teachers committed to disrupting microlevel injustice, slow but deep change is possible. What a vision it will be to one day see people of all kinds regarded as mathematical authorities.

CONCLUSION

Templates to Reflect on the Critical Moments

You do not need to respond to these questions in the order in which they appear.

DISRUPTING INJUSTICE: CONVERSATIONS TO CHANGE OUR PRACTICE

Tandrea Fulton's Critical Moments Reflection

Section 1: Analyzing a Narrative

1. Where did Tandrea experience internal conflict and external conflict?

Section 2: Weigh the Balance

2. What are the pros and cons of accelerating a student facing Malik's circumstances?

3. Read the research section of the chapter that connects to the larger issues in education. What is at stake in choosing not to engage or respond?

Section 4: Consider Your Context

4. Consider how your context may differ from Tandrea's.

	TANDREA	YOU
Race	Black	
Gender	Female	
Education	Masters in K–8 math and science education, read articles about opportunity gap	
Years of teaching	9	
School demographics	Black, white, Hispanic. 70% of students receive free and reduced lunch	
Teaching context	Taught both regular and accelerated levels	

5. Have you encountered similar situations in your practice? How were they similar?

____

____

____

____

____

____

6. How might the differences between you and the author change how you respond?

____

____

____

____

____

____

Section 4: Authentically Disrupt Injustice

7. Responding to injustice in mathematics education often requires teachers to be vulnerable and brave. What are some risks you might need to take to provide acceleration opportunities for students at the bottom of the perceived hierarchy of competence?

8. What resources provided in the chapter could build new learning that will help you?

DISRUPTING INJUSTICE: CONVERSATIONS TO CHANGE OUR PRACTICE

Esther Song's Critical Moments Reflection

Section 1: Analyzing a Narrative

1. What aspects of Esther's narrative resonated with you?

2. What fears did Esther share when considering if and how she should respond to what Jordan had written?

Section 2: Weigh the Balance

3. Read Lateefah's research section of the chapter that connects to the larger issues in education. What is at stake if a teacher chooses to not engage or respond?

4. What fears or challenges do you foresee if you needed to respond to a similar situation? What work do you need to do to overcome those fears and challenges?

Section 3: Consider Your Context

5. Consider how your context may differ from Esther's.

	ESTHER	YOU
Race	Asian American	
School	Majority people of color, selective enrollment	
Education	Learned about social justice math education from an expert in the field	
Years of teaching	6	

6. What assets can you uniquely bring to a similar situation? (E.g., personality, knowledge of the community, strong relationship with administration, race, class, access to university partners, etc.)

Section 4: Authentically Disrupt Injustice

7. When reading this account, what (if anything) felt risky about Esther's response? Where can you find space to take a calculated risk when responding to a similar situation?

8. What resources provided in the chapter could build new learning that will help you?

DISRUPTING INJUSTICE: CONVERSATIONS TO CHANGE OUR PRACTICE

Emily Bogusch's Critical Moments Reflection

Section 1: Analyzing a Narrative

1. Where did Emily experience conflict, internal and/or external?

2. When could Emily have chosen to disengage with her colleague?

Section 2: Weigh the Balance

3. What prevents you from addressing similar moments? What negative consequences could result?

____

____

____

____

____

____

4. To what extent can a student's experience in mathematics education be affected by convincing just one teacher of a perspective shift?

____

____

____

____

____

____

5. Read the research section of the chapter that connects to the larger issues in education. What is at stake in choosing not to engage or respond?

____

____

____

____

____

____

Section 3: Consider Your Context

Consider how your context may differ from Emily's.

	EMILY	YOU
Race	white	
Setting	Off-site professional development workshop	
Social justice background	Social justice focus began in high school	
Years of teaching	12	

6. How might the differences between your and Emily's context change how you would respond to a fellow math educator who has a deficit perspective?

7. What assets can you uniquely bring to the situation? (E.g., personality, knowledge of the community, class, education, etc.)

Section 4: Authentically Disrupt Injustice

8. Which recommendations for action most resonate with you?

9. Which are most challenging? What steps will you need to take to overcome the perceived challenges?

10. What resources provided in the chapter could build new learning that will help you?

DISRUPTING INJUSTICE: CONVERSATIONS TO CHANGE OUR PRACTICE

Evan Taylors's Critical Moments Reflection

Section 1: Analyzing a Narrative

1. What aspects of the narrative resonated with you?

2. What fears did Evan have to consider when rethinking this teaching practice?

Section 2: Weigh the Balance

3. Read the research section of the chapter that connects to the larger issues in education. How is the unjust structure you oppose not only bad for students from underrepresented backgrounds in mathematics education but also bad for you?

Section 3: Consider Your Context

4. Consider how your context may differ from Evan's.

	EVAN	YOU
Race	Black	
Gender	Male	
District pressure	Acceleration	
Tenure	New to school	
Years of teaching	8	

5. Have you encountered a similar tension in your classroom that started with a student request for an instructional or curricular change?

____

____

____

____

____

6. What aspects of Evan's background or identity are different from your own and would affect how you choose to respond?

____

____

____

____

____

Section 4: Authentically Disrupt Injustice

7. Responding to injustice in mathematics education often requires teachers to be vulnerable and brave. What are some possible risks that could be taken?

____

____

____

____

____

8. What next steps can you take to listen to your students' needs?

____

____

____

____

____

9. What resources provided in the chapter could build new learning that will help you?

____

____

____

____

____

DISRUPTING INJUSTICE: CONVERSATIONS TO CHANGE OUR PRACTICE

Kendrick Savage's Moments Reflection

Section 1: Analyzing a Narrative

1. Why did Kendrick experience internal conflict from a short conversation?

____

____

____

____

____

____

Section 2: Weigh the Balance

2. Read the research section of the chapter that connects to the larger issues in education. Are there students in your classroom who might feel invisible?

____

____

____

____

____

____

Section 3: Consider Your Context

Consider how your context may differ from Kendrick's.

	KENDRICK	YOU
Race	Black	
Gender	Male	
College history	First-generation college graduate	
Relationship with colleague	Mentor who has been helpful	
Education	Started a doctorate program in mathematics	

3. Can you recall a moment when you wish you had spoken up but chose not to? What were the reasons that prevented you from doing so?

Section 4: Authentically Disrupt Injustice

4. Responding to injustice in mathematics education often requires teachers to be vulnerable and brave. But for many educators of Color like Kendrick, simply being present in a math space is a challenge in and of itself. Are there moments to see yourself with compassion even if you chose not to respond to an injustice? Explain your response.

5. How can you respond to a similar situation that can challenge a coworker to learn while also honoring your own needs to feel emotionally and physically safe?

DISRUPTING INJUSTICE: CONVERSATIONS TO CHANGE OUR PRACTICE

Michelle Lo's Critical Moments Reflection

Section 1: Analyzing a Narrative

1. What do you identify as the main tension(s) Ms. Lo experienced during the incident?

Section 2: Weigh the Balance

2. What fears do you have when addressing recent political events in math class?

____

____

____

____

____

____

3. Read Teresa Dunleavy's research section of the chapter that connects to the larger issues in education. What is your reaction to the claim that math is always political? Who benefits from the belief that math is neutral or apolitical? Who suffers from this belief?

____

____

____

____

____

____

Section 3: Consider Your Context

Consider how your context may differ from Ms. Lo 's.

	MS. LO	YOU
Race	Taiwanese American	
Parents' political beliefs	Avoid political issues and potential conflict	
Education	Master's and profes-sional development learning around math education and social justice	
Years of teaching	1	
School demographics	Racially diverse, Title I	

6. What actions did Ms. Lo take that you wouldn't do yourself? What actions do you think you would take that Ms. Lo didn't?

7. How do your identity/social markers and history inform actions that feel more or less comfortable to take?

Section 4: Authentically Disrupt Injustice

8. How can you prepare political conversations while considering your own comfort and risk tolerance rather than simply avoiding them?

References

Adichie, C. N. (2009). *The danger of a single story* [Video]. TED Conferences. https://www.ted.com/talks/chimamanda_ngozi_adichie_the_danger_of_a_single_story/c

Aguirre, J. M., Mayfield-Ingram, K., & Martin, D. B. (2013). Identities, agency, and mathematical proficiency: What teachers need to know to support student learning. In *The impact of identity in K-8 mathematics learning and teaching: Rethinking equity-based practices*. National Council of Teachers of Mathematics.

Aguirre, J., Mayfield-Ingram, K., & Martin, D. B. (2013). *The impact of identity in K-8 mathematics learning and teaching: Rethinking equity-based practices*. National Council of Teachers of Mathematics.

Ash, S. L., & Clayton, P. H. (2009). Generating, deepening, and documenting learning: The power of critical reflection in applied learning. *Journal of Applied Learning in Higher Education, 1*(1), 25–48. https://hdl.handle.net/1805/4579

Association of State Supervisors of Mathematics, NCSM, & National Council of Teachers of Mathematics (NCTM). (2021). *Continuing the journey: Mathematics learning 2021 and beyond*. https://www.nctm.org/mathematics2021/

Battey, D., & Leyva, L. A. (2016). A framework for understanding whiteness in mathematics education. *Journal of Urban Mathematics Education, 9*(2), 49–80. https://doi.org/10.21423/jume-v9i2a294

Berry, R. Q. (2003). Mathematics standards, cultural styles, and learning preferences: The plight and the promise of African American students. *The Clearing House, 76*(5), 244–249. http://www.jstor.org/stable/30189841

Berry, R. Q., III (2008). Access to upper-level mathematics: The stories of successful African American middle school boys. *Journal for Research in Mathematics Education, 39*(5), 464–488. https://doi.org/10.5951/jresematheduc.39.5.0464

Berry, R. Q., III., Thunder, K., & McClain, O. L. (2011). Counter narratives: Examining the mathematics and racial identities of Black boys who are successful with school mathematics. *Journal of African American Males in Education, 2*(1), 10–23. https://bma.issuelab.org/resources/22930/22930.pdf

Brown, C. S., & Stone, E. A. (2016). Gender stereotypes and discrimination: How sexism impacts development. *Advances in child development and behavior, 50*, 105–133. https://doi.org/10.1016/bs.acdb.2015.11.001

Carson, A. (2013). *Red doc*. McClelland & Stewart.

Castagno, A. E., & Brayboy, B. (2008). Culturally responsive schooling for indigenous youth: A review of the literature. *Review of Educational Research, 78*(4), 941–993. https://doi.org/10.3102/0034654308323036

Center for Restorative Process. Teaching restorative practices with classroom circles. Accessed January 10, 2022. http://www.centerforrestorativeprocess.com/teaching-restorative-practices-with-classroom-circles.html

Chen, G. A., Marshall, S. A., & Horn, I. S. (2021). "How do I choose?": Mathematics teachers' sensemaking about pedagogical responsibility. *Pedagogy, Culture & Society, 29*(3), 379–396. https://doi.org/10.1080/14681366.2020.1735497

Chen, G. A., & Buebuellll, J. Y. (2018). Of models and myths: Asian (Americans) in STEM and the neoliberal racial project. *Race Ethnicity and Education, 21*(5), 607–625. https://doi.org/10.1080/13613324.2017.1377170

Collins, K. H., Joseph, N. M., & Ford, D. Y. (2020). Missing in action: Gifted Black girls in science, technology, engineering, and mathematics. *Gifted Child Today, 43*(1), 55–63. https://doi.org/10.1177/1076217519880593

Copur-Gencturk, Y., Cimpian, J. R., Lubienski, S. T., & Thacker, I. (2019). Teachers' bias against the mathematical ability of female, Black, and Hispanic students. *Educational Researcher, 49*(1), 30–43. https://doi.org/10.3102/0013189X1989

Countryman, J. (1992). *Writing to Learn Mathematics: Strategies that Work, K-12* (1st ed.). Pearson Education Canada.

Daniszewski, J. (2020, July 20). Why we will lowercase white. https://blog.ap.org/announcements/why-we-will-lowercase-white

Davis, J., Anderson, C., & Parker, W. (2019). Identifying and supporting black male students in advanced mathematics courses throughout the K-12 pipeline. *Gifted Child Today, 42*(3), 140–149. https://doi.org/10.1177/1076217519842234

Duncan-Andrade, J. (2009). Note to educators: Hope required when growing roses in concrete. *Harvard Educational Review, 79*(2), 181–194. https://doi.org/10.17763/haer.79.2.nu3436017730384w

Dunleavy, T. K. (2015). Delegating mathematical authority as a means to strive toward equity. *Journal of Urban Mathematics Education, 8*(1), 62–82. https://doi.org/10.21423/jume-v8i1a242

Dunleavy, T. K. (in press). Status, Antiracism, and Mathematics. In M. Winn & T. Winn (Eds.), *Bloomsbury encyclopedia of social justice in education*. Bloomsbury Publishing Inc.

Dweck, C. S. (2007). The perils and promises of praise. *Educational Leadership, 65*(2), 34–39. https://www.ascd.org/el/articles/the-perils-and-promises-of-praise

Ellison, R. (1952). *Invisible Man*. Random House.

Emdin, C. (2021). *Ratchetdemic: Reimagining academic success*. Beacon Press.

Engle, J., & Tinto, V. (2008). Moving beyond access: College for low-income, first-generation students. The Pell Institute. https://files.eric.ed.gov/fulltext/ED504448.pdf

Fleming, C. M. (2021). *Rise Up! How you can join the fight against white supremacy*. Henry Holt and Company.

Flores, A. (2007). Examining disparities in mathematics education: Achievement gap or opportunity gap? *The High School Journal, 91*(1), 29–42. https://doi.org/10.1353/hsj.2007.0021

Foley, D. E. (1996). The cultural production of the silent Indian. In B.A. Levinson, D.E. Foley, & D.C. Holland (Eds.), *The cultural production of the educated person: Critical ethnographies of schooling and local practice* (pp. 79–92). SUNY Press. https://doi.org/10.1525/aa.1998.100.2.575

Ford, D. Y., & Russo, C. J. (2016). Historical and legal overview of special education overrepresentation: Access and equity denied. *Multiple Voices for Ethnically Diverse Exceptional Learners, 16*(1), 50–57. https://doi.org/10.56829/2158-396X.16.1.50

Georgiou, S. N., Christou, C., Stavrinides, P., & Panaoura, G. (2002). Teacher attributions of student failure and teacher behavior toward the failing student. *Psychology in the schools, 39*(5), 583–595. https://doi.org/10.1002/pits.10049

Gilbert, D. T., & Malone, P.S. (1995). The correspondence bias. *Psychological bulletin, 117*(1), 21–38. https://doi.org/10.1037/0033-2909.117.1.21

Goffney, I., Gutiérrez, R., & Boston, M. (2018). *Rehumanizing mathematics for Black, Indigenous, and Latinx students.* National Council of Teachers of Mathematics.

Gonzalez, N., Andrade, R., Civil, M., & Moll, L. (2001). Bridging funds of distributed knowledge: Creating zones of practices in mathematics. *Journal of Education for Students Placed at Risk, 6*(1–2), 115–132. https://doi.org/10.1207/S15327671ESPR0601-2_7

Graham, K., Burrill, G., & Curtis, J. (2018). *Catalyzing change in high school mathematics: Initiating critical conversations.* National Council of Teachers of Mathematics.

Greer, B., & Mukhopadhyay, S. (2003). What is mathematics education for? *The Mathematics Educator, 13*(2), 2–6.

Gutiérrez, R. (2013). The sociopolitical turn in mathematics education. *Journal for Research in Mathematics Education, 44*(1), 37–68. https://doi.org/10.5951/jresematheduc.44.1.0037

Gutiérrez, R. (2016). Strategies for creative insubordination in mathematics teaching. *Special Issue Mathematics Education: Through the Lens of Social Justice, 7*(1), 52–62. https://files.eric.ed.gov/fulltext/ED584302.pdf

Gutstein, E. (2006). *Reading and writing the world with mathematics: Toward a pedagogy for social justice.* Taylor & Francis.

Hall, H. R. (2015). Food for thought: Using critical pedagogy in mentoring African American adolescent males. *The Black Scholar, 45*(3), 39–53. https://doi.org/10.1080/00064246.2015.1049328

Halberstam, J. (2013). The wild beyond: With and for the undercommons. In Harney, S., & Moten, F., (Eds.), *The undercommons: Fugitive planning and black study* (pp. 2–13). Autonomedia.

Hand, V. M. (2010). The co-construction of opposition in a low-track mathematics classroom. *American Educational Research Journal, 47*(1), 97–132. https://doi.org/10.3102/0002831209344216

Haycock, K. (1998). Good teaching matters: How well-qualified teachers can close the gap. The Education Trust. https://assets.pearsonschool.com/asset_mgr/legacy/200727/1999_04Haycok_397_1.pdf

hooks, b. (1994). *Teaching to transgress.* Routledge.

hooks, b. (2004). Understanding patriarchy. http://imaginenoborders.org/pdf/zines/UnderstandingPatriarchy.pdf

Hsiao, K. P. (1992). First generation college students. *Eric digest*, 1–5.

Okun, T. (2001). White Supremacy Culture. In *Dismantling racism: A workbook for social change groups*. Accessed January 9, 2022. https://www.dismantlingracism.org/uploads/4/3/5/7/43579015/okun_-_white_sup_culture.pdf

Joseph, N. M., Hailu, M. F., & Matthews, J. S. (2019). Normalizing Black girls' humanity in mathematics classrooms. *Harvard Educational Review, 89*(1), 132–155. https://doi.org/10.17763/1943-5045-89.1.132

Kendi, I. X. (2019). *How to be an antiracist.* Allyn & Bacon.

Kennedy, M. M. (2011). Attribution error and the quest for teacher quality. *Educational Researcher, 39*(8), 591–598. https://doi.org/10.3102/0013189X10390804

King, M. L., Jr. (1967). *Where do we go from here: Chaos or community?* Harper & Row.

King, T. (2003). *The truth about stories: A native narrative.* House of Anansi.

Kokka, K. (2015). Addressing dilemmas of social justice mathematics through collaboration of students, educators, and researchers. *Educational Considerations, 43*(1), 13–21. https://doi.org/10.4148/0146-9282.1031

Kokka, K. (2019). Healing-informed social justice mathematics: Promoting students' sociopolitical consciousness and well-being in mathematics class. *Urban Education, 54*(9), 1179–1209. https://doi.org/10.1177/0042085918806947

Kumashiro, K. K. (2000). Toward a theory of anti-oppressive education. *Review of Educational Research, 70*(1), 25–53. https://doi.org/10.3102/00346543070001025

Ladson-Billings, G. (2007). Pushing past the achievement gap: An essay on the language of deficit. *The Journal of Negro Education, 76*(3), 316–323.

Leonard, J., & Martin, D. B. (Eds.). (2013). *The brilliance of Black children in mathematics.* Information Age Publishing.

Lo, M. (in press). CTRL Z: Undoing and rethinking power and student voice in the classroom. *Kaleidoscope: Educator Voices and Perspectives.*

Lo, M., & Dunleavy, T. (2020). *Using complex instruction to dismantle white supremacy culture in mathematics.* (MET Classroom Research Grant 7–12). National Council of Teachers of Mathematics.

Martin, D. B. (2021). In my opinion: Does race matter? *Mathematics Teacher: Learning and Teaching PK-12, 114*(7), 562–569. https://doi.org/10.5951/MTLT.2021.0075

Martin, D. B. (2009). Researching race in mathematics education. *Teachers College Record, 111*(2), 295–338. https://doi.org/10.1177/016146810911100208

Martin, D. B. (2012). Learning mathematics while Black. *Educational Foundations, 26,* 47–66. https://files.eric.ed.gov/fulltext/EJ968817.pdf

Masta, S. (2018). Strategy and resistance: How Native American students engage in accommodation in mainstream schools. *Anthropology & Education Quarterly, 49*(1), 21–35. https://doi.org/10.1111/aeq.12231

Matthews, L. E., Jones, S. M., & Parker, Y. A. (2022). *Engaging in culturally relevant math Tasks: Fostering hope in the elementary classroom.* Corwin Press.

McFarland, D. A. (2001). Student resistance: How the formal and informal organization of classrooms facilitate everyday forms of student defiance. *The American Journal of Sociology, 10* (3), 612–678. https://ed.stanford.edu/sites/default/files/mcfarland/AJS2001.pdf

McIntosh, P. (2003). White privilege: Unpacking the invisible knapsack. In S. Plous (Ed.), *Understanding prejudice and discrimination* (pp. 191–196). McGraw-Hill.

Mickelson, R. A., Bottia, M. C., & Lambert, R. (2013). Effects of school racial composition on K–12 mathematics outcomes: A metaregression analysis. *Review of Educational Research, 83*(1), 121–158. https://doi.org/10.3102/0034654312475322

Moje, E. B., Ciechanowski, K. M., Kramer, K., Ellis, L., Carrillo, R., & Collazo, T. (2004). Working toward third space in content area literacy: An examination of every-day funds of knowledge and Discourse. *Reading Research Quarterly, 39*(1), 38–70. https://doi.org/10.1598/RRQ.39.1.4

Morrison, T. (1992). *Playing in the Dark: Whiteness and the Literary Imagination.* Harvard University Press.

National Council of Teachers of Mathematics. (2000). *Principles for School Mathematics.* National Council of Teachers of Mathematics.

National Council of Teachers of Mathematics. (2020). *Catalyzing change in elementary school mathematics.* National Council of Teachers of Mathematics.

National Mentoring Resource Center. (n.d.). *Mentoring of youth.* https://nationalmentoringresourcecenter.org/resources/what-is-mentoring/

Okun, T. (2021). *White Supremacy Culture.* White Supremacy Culture. https://www.whitesupremacyculture.info/

Oregon Coalition Against Domestic & Sexual Violence. (2018). *Toolkit for Interrupting Oppression.* https://www.ocadsv.org/sites/default/files/resource_pub/ocadsv_toolkit-for-interrupting-oppression_web.pdf

Oskineegish, M. (2015). Are you providing an education that is worth caring about? Advice to non-native teachers in northern First Nations communities. *Canadian*

Journal of Education, 38(3), 1–25. https://journals.sfu.ca/cje/index.php/cje-rce/article/view/1770/1770

Perry, R. P. (2003). Perceived (academic) control and causal thinking in achievement settings. *Canadian Psychology / Psychologie Canadienne, 44*(4), 312–331. https://doi.org/10.1037/h0086956

Reyna, C., & Weiner, B. (2001). Justice and utility in the classroom: An attributional analysis of the goals of teachers' punishment and intervention strategies. *Journal of Educational Psychology, 93*(2), 309–319. https://doi.org/10.1037/0022-0663.93.2.309.

Ross, L. (1977). The intuitive psychologist and his shortcomings: Distortions in the attribution. In L. Berkowitz, (Ed.), *Advances in Experimental Social Psychology* (pp. 173–220). Academic Press. https://doi.org/10.1016/S0065-2601(08)60357-3

Sánchez, B., Hurd, N. M., Neblett, E. W., & Vaclavik, D. (2018). Mentoring for black male youth: A systematic review of the research. *Adolescent Research Review, 3*(3), 259–278. https://doi.org/10.1007/s40894-017-0074-z

Sassler, S., Glass, J., Levitte, Y., & Michelmore, K. M. (2017). The missing women in STEM? Assessing gender differentials in the factors associated with transition to first jobs. *Social Science Research, 63*, 192–208. https://doi.org/10.1016/j.ssresearch.2016.09.014

Savage, K. (2020). Mathematical motivation beliefs: The impact of teachers and parents on student motivation. *The Negro Educational Review, 71*(1-4), 53–82. https://files.eric.ed.gov/fulltext/EJ1153299.pdf

Singleton, G. E., & Hays, C. (2008). Beginning courageous conversations about race. In M. Pollock (Ed.), *Everyday antiracism: Getting real about race in school* (pp. 18–23).

Singleton, G. E., & Linton, C. E. (2007). *Facilitators guide, courageous conversations about race: A field guide for achieving equity in schools.* Corwin Press.

Sleeter, C. E. (2017). Critical race theory and the whiteness of teacher education. *Urban Education, 52*(2), 155–169. https://doi.org/10.1177/0042085916668957

Sleeter, C. E., & Grant, C. A. (2017). Race, class, gender, and disability in current textbooks. In *The politics of the textbook* (pp. 78–110). Routledge.

Steele, C. M. (1997). A threat in the air: How stereotypes shape intellectual identity and performance. *American Psychologist, 52*(6), 613–629. https://doi.org/10.1037//0003-066x.52.6.613

Steele, C. M., & Aronson, J. (2005). Stereotypes and the fragility of academic competence, motivation, and self-concept. In A. J. Elliot, & C. S. Dweck, (Eds.), *Handbook of competence and motivation* (pp. 436–455). Guilford Publications.

Stiff, L. V. & Johnson, J. L. (2011). Mathematical reasoning and sense making begins with the opportunity to learn. In M. E. Strutchens, & J. R. Quander, (Eds.), *Focus in high school mathematics: Fostering reasoning and sense making for all students* (pp. 85–100). National Council of Teachers of Mathematics.

Sue, D. W., Capodilupo, C. M., Torino, G. C., Bucceri, J. M., Holder, A. M., Nadal, K. L., & Esquilin, M. (2007). Racial microaggressions in everyday life: implications for clinical practice. *American psychologist, 62*(4), 271–286. https://doi.org/10.1037/0003-066X.62.4.271.

Tate, W.F. (2006). Race, retrenchment, and the reform of school mathematics. In E. Gustein, & B. Peterson (Eds.), *Rethinking mathematics: Teaching social justice by the numbers* (pp. 3–42). Rethinking Schools, Ltd.

Thomas-Alexander, S., & Harper, B. E. (2017). Cleaning up the clinic: Examining mentor teachers' perceptions of urban classrooms and culturally responsive teaching. *Multicultural Learning and Teaching, 12*(1), 49–65. https://doi.org/10.1515/mlt-2015-0013.

TODOS: Mathematics for ALL, & NCSM. (2016). *Mathematics education through the lens of social justice: Acknowledgment, actions, and accountability.* https://www.todos-math.org/assets/docs2016/2016Enews/3.pospaper16_wtodos_8pp.pdf

Valencia, R. R., & Black, M. S. (2002). "Mexican Americans don't value education!"—On the basis of the myth, mythmaking, and debunking. *Journal of Latinos and Education, 1*(2), 81–103. https://www.sjsu.edu/people/marcos.pizarro/courses/215/s0/ValenciaBlack.pdf

Villegas, A. M., & Lucas, T. (2002). Preparing culturally responsive teachers: Rethinking the curriculum. *Journal of Teacher Education, 53*(1), 20–32. https://doi.org/10.1177/0022487102053001003

Villenas, S. (1996). The colonizer/colonized Chicana ethnographer: Identity, marginalization, and co-optation in the field. *Harvard Educational Review, 66*(4), 711–732. https://doi.org/10.17763/haer.66.4.3483672630865482

Whiting, G. W., & Ford, D. Y. (2009). Multicultural issues: Black students and advanced placement classes: Summary, concerns, and recommendations. *Gifted Child Today, 32*(1), 23–26. https://doi.org/10.4219/gct-2009-840

Woodward, B., & Howard, T. C. (2015). Creating educational excellence in the pipeline: Undergraduate African American males' perceptions of educational research and participation in the Black male project. *Spectrum: A Journal on Black Men, 4*(1), 99–125. https://doi.org/10.2979/SPECTRUM.4.1.06

Yu, X. (2021). Everybody still plays: Virtual engagement without webcams on. *Mathematics Teacher: Learning and Teaching PK-12, 114*(10), 759–767. https://doi.org/10.5951/MTLT.2021.0006